DAVID O. MCKAY LIBRARY
P9-CJE-714

Intelligent Design or Evolution?

Why the Origin of Life and the Evolution of Molecular Knowledge Imply Design

By Stuart Pullen

OCT 3 1 2005

WITHDRAWN
JUN 0 5 2024
DAVID O. McKAY LIBRARY
BYU-IDAHO

PROPERTY OF:
DAVID O. McKAY LIBRARY
BYU-IDAHO
REXBURG ID 83460-0405

Copyright Information:

Library of Congress Control Number 2005903070

The content of this book was published online in November 2004, http://www.evolution-by-design.com. Readers with high speed internet access should visit this site to view the interactive java applets. These applets use the Jmol molecular viewer to display biological molecules. Several scripts are available to color these molecules by information and knowledge.

© 2005 by Stuart Pullen

Intelligent Design Books
Raleigh, NC

Table of Contents:

About the Author:

The author graduated in 1987 from North Carolina State University with a degree in biochemistry. He then worked for 5 years in this field while pursuing a graduate degree in electrical engineering (1992 also from North Carolina State). The author currently designs integrated circuits used in switching power supplies and holds 15 patents related to this field.

The two degrees proved invaluable in writing this book. Information theory has historically been a discipline confined to electrical engineering, but with the explosion of biological information over the past two decades, information theory is now essential to the biological sciences. The emerging field of bioinformatics combines molecular biology with information theory.

Several fields of science are required to fully characterize the origin of life and its difficulties. Information theory is just one of these. Organic chemistry and chemical thermodynamics are equally important. The author's background allows this publication to investigate the mystery of life's origin from several important perspectives.

Introduction: Evolution vs. Design

This book will show that naturalistic laws do not explain the origin of life and then suggest that this failure implies that life was created. The logic behind this conclusion is known as intelligent design.

The Philosophy of Science

By definition, science must explain everything in terms of naturalistic laws. So before analyzing any data, science rules out the possibility that life was created. While this philosophy has served science well, it is somewhat problematic if a creator exists. Intelligent design differs from science because it does not use assumptions to eliminate possibilities. Instead intelligent design allows the evidence to lead where it may.

Scientific experiments test how evolution happens. Experiments designed to test if evolution is possible are not necessary because science assumes that chance and natural selection are responsible. This assumption has trapped science. To better understand the trap, suppose tomorrow that a thousand fossils are found that document how T. Rex evolved from another dinosaur. Science will assert that these fossils prove that T. Rex evolved from another ancient dinosaur. Further, it will assert that naturalistic laws fully explain the transition. Science does not have to justify the second assertion because it simply assumes that it is true. Furthermore, most scientists would consider any experiment designed to test the probabilities associated with this particular evolutionary transition as unnecessary. This philosophy does not allow science much freedom. No matter what the evidence shows, the assumptions on which science is based ensure that it will always support the theory of evolution. Thus, science is trapped.

The failure of science to ask is evolution possible has led to the premature acceptance of many ideas concerning evolution. For example, when Darwin proposed the theory of evolution in 1859, the chemistry of living organisms was a complete mystery, so the testing of Darwin's theory was limited. For nearly 100 years following Darwin's first publication, science only had three ways to test his theory: 1) search for the fossils that link existing animals and plants, 2) design experiments to observe how animals and plants change ever so slightly from one generation to the next and 3) accumulate evidence that the earth is very old. While such experiments were critical for evolution's acceptance, the chemical processes behind evolution remained a mystery. Science just assumed that naturalistic laws were responsible; thus, science was able to embrace the theory of evolution without understanding it.

The Molecular Theory of Evolution

In 1953, scientists began to unravel the chemistry of life when Watson and Crick proposed a model for DNA. Soon thereafter the genetic code was broken, and the chemical mechanism behind evolution became clear. The hypothesis put forth is outlined below:

Sections of DNA called genes store the information needed to make proteins, and this information is passed from one generation to the next when genes are replicated during reproduction. The replication process is not perfect, and as such it may by chance introduce errors. Errors during replication, mutations, have the potential to create new genes. Mutations may create new information, or they may simply alter existing information. In either case, nature preserves beneficial mutations through the process of natural selection and other mutations survive by chance. Over many millions of years, changes in existing genes yield new genes; therefore, animals continually evolve and adapt.

Soon after its proposal, this hypothesis became the framework for the theory of molecular evolution. While scientists have modified it over the years, the basic framework of the theory remains intact with one important exception.

If an existing gene evolves into a new gene with a new function, then the original function will be lost, and natural selection will not allow this to happen. So Ohno suggested that existing genes do not evolve into new genes unless they are first duplicated.[10] The duplicate copy is free to evolve a new function while the original maintains its current function. Others have refined the theory further by suggesting that pieces of existing genes may be duplicated and then rearranged to create new genes with new functions. With these modifications, the molecular theory certainly explains the origin of many genes.

But even with these improvements, the concern raised earlier remains the same - why not ask if evolution can happen? Science describes how it happens, but why not take the next step and investigate the probabilities associated with the required events. That is rather than assume that naturalistic laws are responsible, prove that these laws are responsible. This avoids the trap. Thus, experiments are needed to test whether or not evolution is possible.

A ten year experiment can hardly hope to model a billion years of evolution, but today there is a solution to this problem. Scientists around the world are actively sequencing the DNA of many animals, plants and bacteria, and after more than three decades of characterization, this information is freely available in online databases. These databases allow science to ask for the very first time two important questions. Can mutations operating over billions of years and guided by natural selection create new genes? And perhaps more importantly, are naturalistic laws responsible?

Is Evolution Possible?

To test if evolution is possible, consider a gene that is common to all living things. That is the gene is found in algae, bacteria, oak trees, carrots, mice, fish, people and all other living things. This gene will not be identical in all living things, but it will be similar. The form of the gene found in bacteria has had at least 2 and probably 3 billion years to change independently from the same gene found in animals and plants. The form of the gene in an oak tree has had at least 1 billion years to evolve independently from the gene in man. This gene and many others like it allow science to conduct experiments that look back in time - almost to the origin of life itself.

A comparison of this ancient gene in many different species will reveal how much information the gene contains, and from such an analysis, one can calculate the probability associated with the gene's evolution. If the probability so calculated is for all practical purposes zero, then design may be inferred. Unfortunately, this inference is not compelling because it fails to consider the effect of natural selection. In order to make the argument for design stronger, the concept of information must be replaced with another familiar concept, knowledge. The two definitions that follow are important.

Molecular information is the information found in a gene today. It is calculated by comparing the differences found in the same gene in many different animals, plants and bacteria. Information has a precise mathematical definition as defined by information theory.

Molecular Knowledge is the minimum amount of useful information required by a gene to have any function. Any region of DNA that does not contain molecular knowledge has no function at all. Thus, such a region cannot be preserved or optimized by natural selection, and it cannot be classified as a gene.

Molecular knowledge is always less than molecular information. Molecular knowledge is more difficult to calculate because it does not have a mathematical definition.

The Evolution of Molecular Knowledge

Because of natural selection, information cannot be used to calculate the probability that a gene will evolve. Information is useful because it has a precise mathematical definition not because it can answer questions concerning whether or not the evolution of a new gene is possible. Chance is not in control if natural selection is guiding which mutations survive. Therefore, relating the amount of information in a gene to a probability that it can evolve is not a valid mathematical analysis.

Molecular knowledge is the minimum amount of useful information required for a gene to have any function. If a gene does not contain molecular knowledge, then it has no function, it confers no selective advantage, and it is not a gene. Thus, before a region of DNA contains the requisite molecular knowledge, natural selection plays no role in guiding its evolution. Chance controls which mutations survive. Thus, molecular knowledge can be related to a probability of evolution. Figure 1 helps illustrate these important concepts.

Notice that the first step that creates the required molecular knowledge is vertical, and the subsequent step that creates molecular information is sloped. This difference is important in that it is meant to show that natural selection can help guide the last transition, but it plays no role in the first. Thus, it is the size of the first step that determines whether or not a gene can evolve.

Figure 1: Information in Life

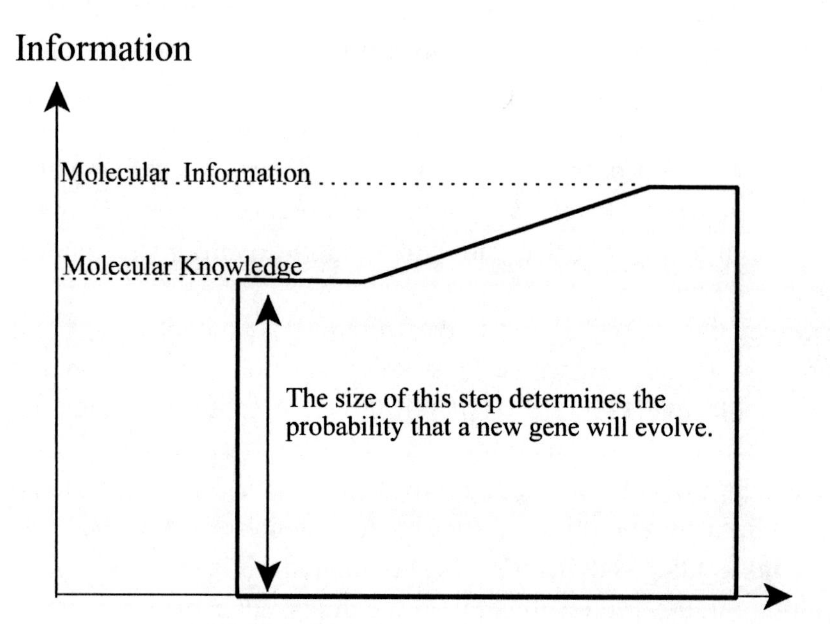

A simple example will now be introduced to help clarify this concept. Consider the following sentences:

I have a 13 year old black lab who likes to fetch a tennis ball. His name is Bubba.

My 13 year old black lab, Bubba, likes to fetch a tennis ball.

mi 13 yr od blk lab, buba, like fetch tenis bal.

Each sentence represents a gene. The first sentence uses the most letters; therefore, by definition, it contains the most information. The second sentence uses fewer letters, communicates the same points and is still grammatically correct. The last sentence uses the fewest letters.

If these sentences are composed by randomly selecting letters from the alphabet, then the probability of spelling the last sentence is much better than the first. So only the last sentence and all sentences similar to it are useful for assigning a probability to the evolution of a gene. In this example, the first sentence represents molecular information and the last one represents molecular knowledge. If some concepts are not required, the last sentence may be simplified further. For example, the sentence fragment, mi black lab, still communicates some knowledge. Thus, finding the precise threshold of molecular knowledge can never be exact. It relies on both math and human insight.

At least one biologist, Richard Dawkins, has suggested that sentences like: "hh n swd dwqdoe ffnfnriiq jddk" still confer a selective advantage; therefore, given time these sentences will evolve into something useful under the guidance of natural selection.[1,14] Dawkins ran many computer simulations with sentences like the one above, and they all evolved into the desired result. But his programming and logic are both flawed because natural selection cannot preserve or optimize a gene that offers no selective advantage (the nonsense sentence above represents a gene that confers no selective advantage). A gene must contain some useful information before natural selection activates. Thus, chance and chance alone must create the initial knowledge. Because the probability that chance can accomplish this goal is proportional to the step height in figure 1, the size of this step completely determines whether or not a new gene will evolve. It is the goal of this book to characterize the size of this initial step. If it is small then naturalistic laws explain the evolution of knowledge. On the other hand, as the step size increases, the probability that chance will create the required molecular knowledge approaches zero, and at some critical threshold, the design inference becomes valid.

Large steps are associated with the evolution of genes that are completely different from all other existing genes. For these genes, the probability of chance finding an appropriate solution (even given 50,000 billion years) is very close to zero. The origin of these genes imply design.

Chemical Evolution

This book will also evaluate another hypothesis put forth in the 19th century. This hypothesis attempts to explain the origin of life and its basic premise is as follows:

The early earth's atmosphere was different from today in that no free oxygen was available. Under these circumstance, energy sources like sunlight and electrostatic discharges might create the chemicals necessary for life (chemical evolution). As these chemicals were concentrated in a small pond or puddle, the primordial soup, they organized themselves in such a way to form the first living organism. Because life is very complex, the first living thing is usually assumed to be a self replicating chemical rather than a living cell. Because the first living thing was able to replicate itself, it evolved into life as it exists today.

This hypothesis or some form of it is found in almost all biology books where it is put forth as the generally accepted theory. Yet in the scientific journals, scientists routinely dismiss many aspects of the hypothesis as highly improbable (Shapiro 1995 and 1999; Miller 1995 and 1998; Joyce 1984 and 1989; Nissenbaum 1975; Ferris 1987; Joyce and Orgel 1999; Thaxton:1984). When it comes to chemical evolution and the origin of life, science just does not have the answer.

One of the first experiments concerning the origin of life was conducted in 1953 by Stanley Miller. Miller created several amino acids (the building blocks that life uses to make proteins) in an electrostatic discharge chamber. The experiments conducted since Miller have demonstrated how difficult it is to create the biological precursors required for life. While several amino acids can be created under plausible conditions, proteins cannot be. Furthermore, DNA is much more problematic because its building blocks are difficult to create. Many of these building blocks are unstable and decay rapidly. Science has yet to offer a plausible explanation for how these hard to make and easy to destroy chemicals accumulated in the primordial soup (see references 3,4,5, 6, 7, 8 and 13 on page 15, and chapter 9).

The most prevalent myth concerning chemical evolution suggests that a continuous flow of energy through a complex system of nonliving chemicals will promote the formation of biologically relevant molecules. The researchers who hold to these views suggest that life arose spontaneously when these biological precursors combined in a small pond or puddle several billion years ago. While such energy flows are critical to the survival of life today, it is not clear how they solve the mystery of life's origin. Life knows how to use these energy flows to do work. Such knowledge is completely lacking from a system that only contains nonliving chemicals. Plentiful energy sources if anything do more harm than good. Sunlight bombarding a small pond on the earth 4 billions years ago is much more likely to destroy any useful biological molecules than create one (see for example Fox, Molecular Evolution and the Origin of Life, p37).

Surprisingly, such difficulties are often overlooked; as a result, many biologists mistakenly believe that it is quite easy to synthesize all of the required biological molecules. Nevertheless, a quick review of the relevant literature reveals that this is not true. For example, to synthesize adenine (one of the most important chemicals found in DNA and RNA), chemists start with a concentrated solution of hydrogen cyanide and ammonia. Concentrating ammonia is not an easy task since it is a gas that boils at sub-freezing temperatures, and it also decays rapidly in the presence of sunlight. Furthermore, concentrating hydrogen cyanide in the presence of water is impossible because it reacts with water quite readily yielding formic acid. Scientists tend to focus on the fact that adenine can be synthesized in a laboratory and ignore the fact that the conditions required for its synthesis did not exist on the primitive earth. [3,4,6]

After 50 years of investigation no plausible prebiotic path exists to synthesize cytosine, ribose or deoxyribose (three critical subunits of DNA and RNA). The problems with ribose and cytosine synthesis are so severe that Miller and several others have suggested that the first self replicating molecule probably contained neither.[7, 8]

Biological molecules may contain thousands of subunits all linked together by chemical bonds. Coercing the subunits to form a large biological molecule like DNA or RNA is not easy. These problems are often discussed in scientific journals like Nature, Science, PNAS, and the Journal of Molecular Evolution. For example, even today, investigators have yet to identify a plausible prebiotic method to link cytosine, thymine or uracil to ribose (a step necessary for DNA and RNA synthesis).[11] Nevertheless, not finding the answer is not news. So only the scientists who read these journals are aware of the difficulties involved.

Finally, the greatest challenge to the origin of life lies not with creating the chemical precursors, but instead with creating the required knowledge. The chemicals that make up life contain useful information, and it is this knowledge that allows life to propagate. The implication is that even if a few of the biological precursors required for life existed in the primordial soup, such precursors would not contain the knowledge necessary to live and evolve.

Joyce and Orgel sum up the situation best "After dreaming of self-replicating ribozymes emerging from pools of random polynucleotides, and having nightmares about the difficulties that must have been overcome for RNA replication to occur in a realistic prebiotic soup, we awaken to the cold light of day . . . It must be said that the details of this process remain obscure and are not likely to be known in the near future." - The RNA World, p72-73.

The Origin of Life

Before trying to understand the hurdles associated with the origin of life, it is useful to define life. In its simplest terms, life is a group of chemicals that possess molecular knowledge. The word *knowledge* implies that the information possessed by the chemicals is useful unlike information which may or may not be useful. The word *molecular* indicates that the knowledge resides in a chemical molecule instead of in a book or some other source.

It is this molecular knowledge that allows the chemicals in life to maintain a state that is very different from nonliving chemicals like vinegar, ammonia, and water. The molecular knowledge that life possesses is both procedural and conditional. Procedural knowledge is knowledge about how to do something. For example, how to extract energy from a sugar molecule and use it to build something else. Conditional knowledge is knowledge about why and when something needs to be done. For example, when there is no sugar present certain metabolic pathways should be turned off. Conditional knowledge in molecules is similar to that found in computer programs. A computer program may execute one command if a certain condition is true and another command if the condition is false. Computers do not think. The decisions are predetermined by the logic used in the computer's code.

It is now possible to develop a concise and accurate definition for life: *Life is a system of chemicals possessing molecular knowledge and a mechanism to implement this knowledge in such a way that the system can survive long enough to replicate itself.*

Today, life requires several chemicals to survive, grow, and reproduce. Two chemicals, DNA and RNA, store the required knowledge. Proteins and to lesser extent RNA implement this knowledge, and a third chemical, ATP, provides the energy to power the implementation. At a minimum, the simplest living system must be able to perform four critical functions:

- Store molecular knowledge.
- Implement this knowledge.
- Tap a plentiful energy source to power the implementation.
- Synthesize any biological molecules required for replication that are not plentiful in the primordial soup.

Herein, lies the mystery behind life's origin. The origin of life is a classic example of the chicken or the egg paradox because none of the critical functions listed above can exist without the others.

Many investigators have tried to overcome the paradox by suggesting that the first living thing was a single chemical that contained both the knowledge and the ability to implement the knowledge. RNA is a natural choice for the first living chemical because it can both store and implement knowledge. Nevertheless, after 25 years of experiments, the RNA hypothesis has yet to live up to its expectations. RNA has quite a bit of trouble with self replication (see Joyce:1989 and chapter 10).

Investigators have for the most part over looked the third critical function required for life, the need to tap an energy source to drive replication. Without this function, self replicating molecules become a special type of perpetual motion machine. A perpetual motion machine is a machine that runs forever with no energy input. Perpetual motion machines do not exist. They may run for a short time, but without a continuous input of energy, they eventually stop. Furthermore, all machines must know how to tap an energy source. A car with an empty gas tank cannot be driven to the gas station just because the sun is out. The sun provides an almost unlimited source of energy, but a gas engine does not know how to convert this energy into work. The same constraints apply to a self replicating RNA molecule. Unless such a molecule knows how to tap a plentiful energy source to drive its own replication, it can only exist in text books and in the imagination of researchers.

To summarize, life requires some minimal molecular knowledge to replicate. This knowledge can be possessed by a single chemical, or it can be spread out among many. In either case, the system must possess the knowledge to replicate, a way to implement this knowledge, and a way to power the implementation. A system that does not possess all three is not a living system. Furthermore, any system that is unable to synthesize the chemicals that are required for replication is not robust. These systems cannot evolve because they cannot self replicate.

Figure 2 depicts the focus of this book. The following chapters will concentrate on the genes and proteins that were required for the origin of life and on the chemicals that gave rise to these first genes and proteins. These are events that happened more than 2 billion years ago.

If a new gene evolves early in life's history, and it is completely different from any other existing gene, then the possibility that it arose by gene duplication can be eliminated. This makes the analysis much more manageable. Furthermore, the techniques used in the book simply will not work to prove that man did not evolve from apes. The DNA in a chimpanzee is almost identical to that of man's DNA. This similarity makes it difficult to infer design.

In figure 2, the self replicating molecule leads to a perpetual motion machine. The car in this figure that is pulled along by the powerful magnet is just one example of such a machine. Since perpetual motion machines do not exist, this pathway is not a very promising solution to the mystery of life's origin. It is far more likely that life arose all at once.

The difficulties associated with chemical evolution suggest that the biological precursors necessary for life would have been scarce if they existed at all, and this scarcity suggests that the first living thing was able to synthesize all of the chemicals that it needed for replication and drive this synthesis with a plentiful energy source. Today, life can only tap plentiful energy sources with the help of proteins and lipids, and this suggests that the first living thing was probably also able to synthesize proteins and lipids. Therefore, the first living thing was not a simple self replicating chemical, but rather a living cell very similar to life as it exists today.

While the idea that life arose all at once is not a popular one as it is contrary to Darwinian evolution, the evidence suggests that it did.

Notice in figure 2 that the Cambrian explosion may also imply design (Meyer: 2005). The fossil record indicates that almost every major biological classification (phylum) arose in a very short time span about 500 million years ago. The question mark is meant to show that any design inference based on the Cambrian explosion is subjective because no scientist has yet to accurately model the probabilities of such an event.

Figure 2: The History of Evolution

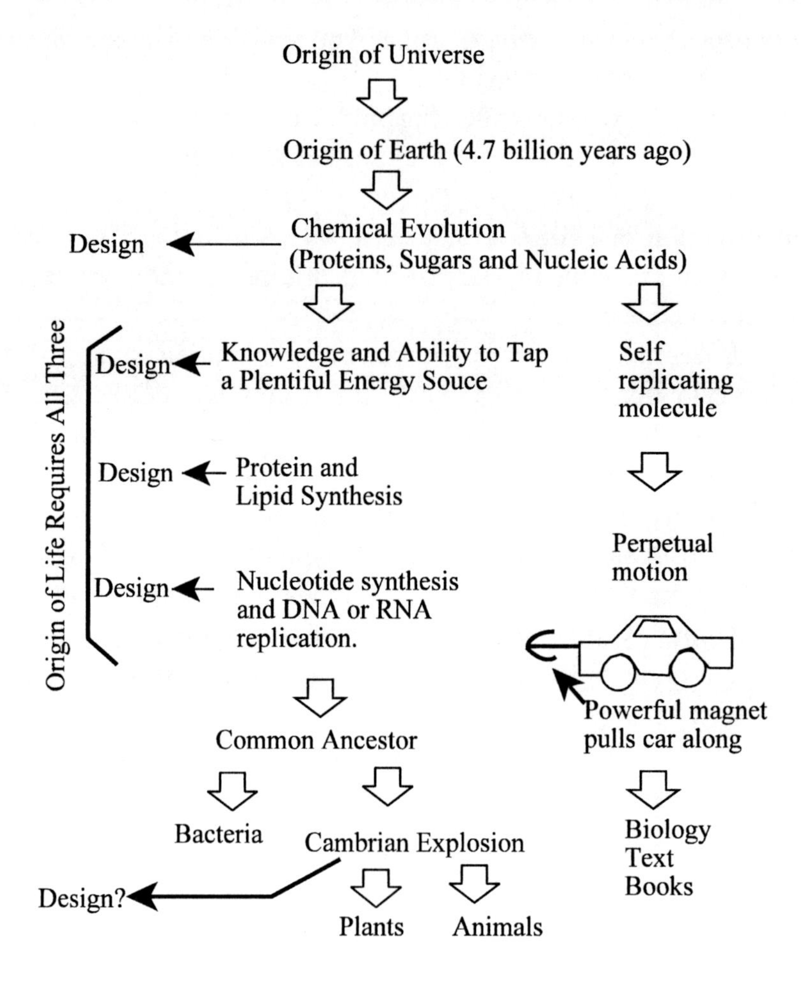

Before proceeding, one final clarification is in order. A few people have intentionally confused creation science with intelligent design. Creation scientists believe that the creation story in Genesis is scientifically accurate. Intelligent design differs from creation science in several important ways. First, intelligent design makes no assumptions as to what the scientific data should show. Intelligent design is just a methodology that uses indirect logic to interpret the scientific evidence. It does not depend on religion. Furthermore, the statistical techniques used by intelligent design require an old earth, common ancestry, and descent with modification.

References:

1) Richard Dawkins, The Blind Watchmaker, Norton and Company, 1996.
2) Fox, Dose, Molecular Evolution and the Origin of Life, Freeman and Company, 1972.
3) Shapiro, "The Prebiotic Role of Adenine: A Critical Analysis", Origins of Life and the Evolution of the Biosphere, 25: 83-98, 1995.
4) Thaxton, Bradley, Olsen, The Mystery of Life's Origin: Reassessing Current Theories, Philosophical Library, 1984.
5) Levy, Miller, "The Stability of the RNA bases: Implications for the Origin of Life," 95: 7933-7938, PNAS, 1998.
6) Shapiro, "Prebiotic Cytosine Synthesis: A Critical Analysis and Implications for the Origin of Life," 96: 4396-4401, PNAS, 1999.
7) Larralde, Robertson, Miller, "Rates of decomposition of Ribose and other Sugars: Implications for chemical Evolution," 92:8158-8160, PNAS, 1995.
8) Joyce, Schwartz, Miller, Orgel, "The Case for an Ancestral Genetic System Involving Simple Analogues of the Nucleotides," PNAS, 84:4398-4402, 1987.
9) Joyce and Orgel, The RNA World, Gesteland, Cech, Atkins, Cold Spring Harbor, "Prospects for Understanding the Origins of the RNA World," 1999. 10) Ohno, Evolution by Gene Duplication, Springer Verlag, 1970.
11) Fuller, Sanchez, Orgel, "Solid state Synthesis of Purine Nucleotides," Journal of Molecular Evolution, 1975.
12) Meyer, "The Origin of Biological Information and the Higher Taxonomic Categories," Proc. of the Biological Society of Washington, 117: 213-239, 2005.
13) Miller, "Which Organic Compounds Could Have Occurred on the Prebiotic Earth?", Cold Spring Harbor Symposia on Quantitative Biology, Volume L11, 17-25, 1987.
14) Spetner, Not By Chance! Shattering the modern Theory of Evolution, 1997.

Part 1: The Evolution of Knowledge and Information

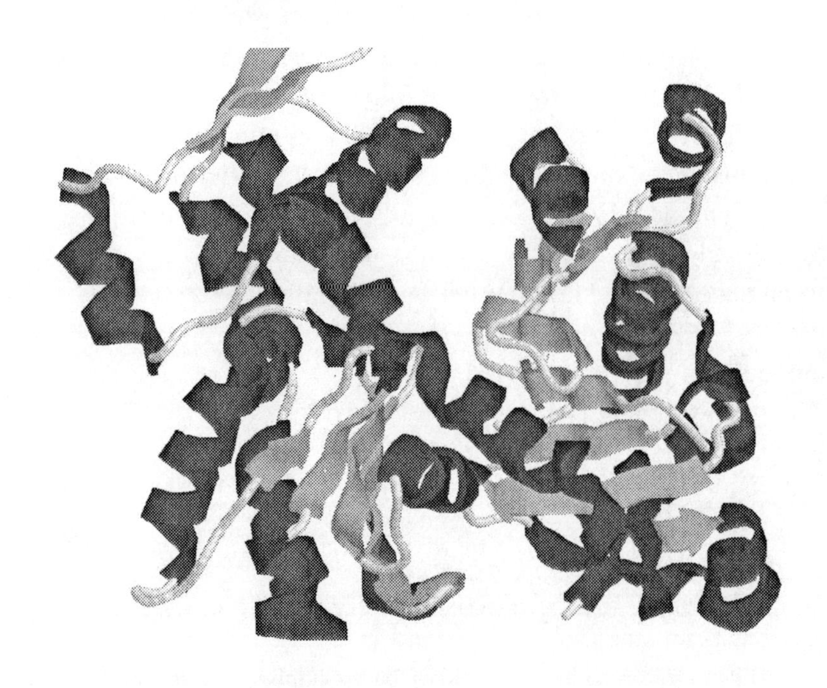

Cartoon representation of the crystal structure of the protein, actin. Actin is one of the proteins required for muscle contraction.

Chapter 1: Information vs. Knowledge

One of the goals in this book is to investigate how molecular knowledge evolves in biological systems. Because there is no mathematical definition for knowledge, much of this investigation will be open to interpretation. To find the molecular knowledge possessed by a chemical like DNA, it is first necessary to quantify the amount of information. This is true because unlike knowledge, information has a precise mathematical definition.

The scientific definition for information is very different from the common one. The everyday definition implies that information should be useful or at least convey some amount of knowledge. The scientific definition does not make any distinction between useful and useless information. For example, consider the following sentences:

The brown dog likes to fetch a tennis ball.

Zxrd zgbzbue awfllt jhjzhwzhg zwnzi oppwnnni wyxaz.

If information theory is applied to these two sentences, the results will indicate that the second sentence contains much more information than the first. Not only is the second sentence longer, but it contains many letters that are rarely used in English (z, x, and w). The first sentence contains useful information. The second message contains no useful information. Yet information theory asserts that the second contains more information than the first. How can this be?

To understand why, consider why scientists developed information theory. The theory was developed by an engineer, Claude Shannon, who was interesting in transmitting information. The second sentence takes longer to transmit than the first, so it contains more information. This definition is clearly not useful to biologists studying evolution.

Evolution involves the creation of information that provides a selective advantage. That is the organism that possesses the new information has an edge over those that do not. Therefore, the information must be useful. This is why the word knowledge is preferable. Knowledge implies that information is useful.

In communication systems, information does not have to contain knowledge. In general, the same cannot be said for biological systems. Information that does not provide a selective advantage is often lost. Thus, the information found in biological systems usually conveys knowledge, and this knowledge provides a selective advantage. In biological systems, knowledge and information are often related, but they are not necessarily equal. Consider the following two sentences:

I have a dog. His name is Bubba. He is a black lab. He is 13 years old.

My black lab, Bubba, is 13.

Both sentences describe four identical concepts, so the knowledge conveyed by both is identical, but the first sentence contains much more information than the second. Because information has a precise mathematical definition, it can be determined rather easily. In contrast, knowledge will always be open to interpretation. Nevertheless, it is possible to define molecular knowledge in terms of information. The proposed definition is as follows:

Molecular Knowledge: the minimum amount of information necessary to enable a chemical (or group of chemicals) to accomplish some task or to specify some trait. The only stringent requirement is that molecular knowledge must confer a selective advantage so that natural selection can preserve it.

Because molecular knowledge is now defined in terms of information, information theory can be used in conjunction with human insight to calculate knowledge. The rest of this chapter will explore information and its properties.

The Nature of Information

Mathematically, information is defined as a reduction in uncertainty. Consider a scientist trapped in a room. He has a coin and a telephone. He is told to flip the coin and then tell his colleagues who are 500 miles away the results using the telephone. He is to repeat this process until he is told otherwise.

Before the scientist flips the coin, he does not know whether it will land heads or tails. There are two possible outcomes, and the scientist does not know which will happen until he observes the results. Suppose that the first toss is heads. As soon as the scientist observes this result, he has information. Two possibilities have been reduced to one. Before observing the coin, the scientist was uncertain of the outcome. After he observes the result, he is certain of the outcome. His colleagues do not have any information until he tells them that the coin landed heads.

A unit of information is called a bit. Whenever two possible outcomes are reduced to one, one bit of information is created. Thus, the scientist acquires one bit of information each time he tosses the coin and observes the result.

Uncertainty depends on the number of possible outcomes. For example, if the scientist is given a die to roll instead of a coin his uncertainty increases. With the coin, there are only two possible outcomes, and with a die there are six. The reduction in uncertainty for a die (six possible outcomes reduced to one outcome) is greater than it is for the coin (two possible outcomes reduced to one outcome).

Thus, the scientist acquires more information when he observes the results of tossing the die than when he observes the results of tossing the coin.

Trapped Scientist with Three Coins

Consider a scientist trapped in a room. He is given three coins, instructed to toss the coins one at a time, and enter the results into the computer (figure 1.1). If the coin lands tails, he is instructed to enter the letter *T*, and if it lands heads, he is instructed to enter the letter *H*. The combination for the door is H-T-H. The scientist has a 1 in 8 chance of opening the door.

Figure 1.1: Trapped Scientist with Three Coins

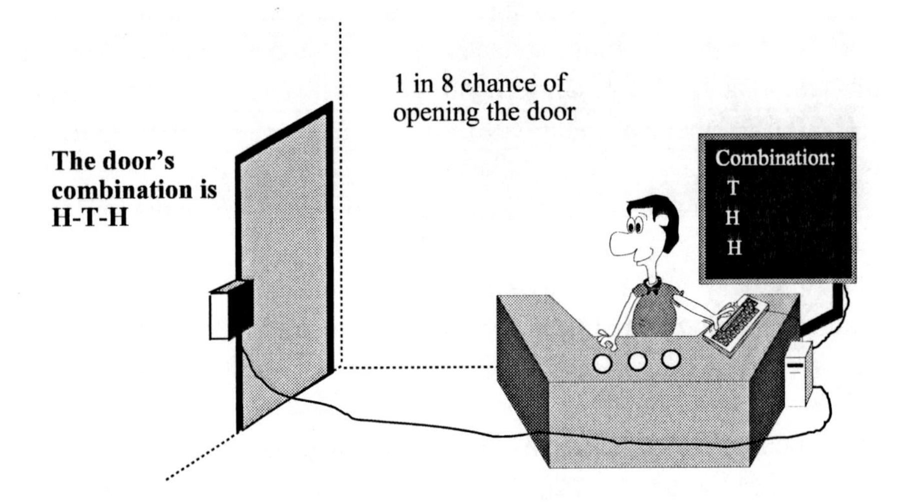

When three coins are tossed, there are eight possible outcomes:

Coin 1	Coin 2	Coin 3	
H	H	H	
H	H	T	
H	**T**	**H**	--------->opens the door
H	T	T	
T	H	H	
T	H	T	
T	T	H	
T	T	T	

When the first coin is tossed, it has two ways to land, heads or tails. The same rules apply for the second and third. So the total number of possible outcomes is 2x2x2 = 8.

Whenever the scientist tosses a coin and observes the results, he acquires information. When he tosses the first coin and observes the result, he acquires one bit of information. After he observes the result of the second coin, he possesses 2 bits of information, and after the third, he possesses 3 bits. Suppose on his first try to open the door, all three coins land heads. After observing this event, the scientist will possess 3 bits of information. He keeps trying, and after a few more tries, the first coin lands head, the second lands tails and the third lands heads. When he enters this result into the computer, the door opens. The scientist has acquired knowledge. The combination for the door is H-T-H, and he now knows the combination.

Notice that every time the scientist tosses the coin he creates information, but only one specific outcome creates useful information or knowledge.

One bit of information corresponds to each coin. In figure 1.1, all results contain 3 bits of information. One result, H-T-H, contains 3 bits of knowledge.

Suppose that the combination is changed to H-T-H-H-H-H-T-H-H-H-H-H-H-H-H-H-H-H-H-H. The scientist is given 20 coins, told to toss all 20, enter the results into the computer and observe the door. How much information is generated every time the scientist tosses 20 coins and observes the result? Answer: 20 bits because there are 20 coins. While 20 bits of information are generated with each attempt to open the door, only one possible outcome will open the door. This is the only outcome that contains both information and knowledge.

With 20 coins, what is the probability that the scientist will find the correct combination on the first try? Answer: multiply 2 by itself 20 times to determine the total number of possible outcomes (1,048,576). Because only one of these outcomes will open the door, the odds are 1 in 1,048,576 or approximately 1 in a million.

Exponents are a useful shorthand for representing a number multiplied by itself many times. The phrase 2 multiplied by itself 20 times can be written as 2^{20}. The number *10* multiplied by itself 86 times can be written as 10^{86}. See appendix three for a review of exponents.

Information Is Closely Related to Probability

Suppose that the scientist has 100 coins, and he is told that he has a 1 in 64 chance of opening the door if he tosses the correct number of coins and enters an *H* or *T* into the computer. He is told not to toss all 100 coins because the combination for the door is not that long. He is also told that the door only has one correct combination.

How does the scientist figure out how many coins to toss? He needs to convert the odds of opening the door into an equivalent number of bits. One way is trial and error. He can compose a table like table 1.1. If the scientist tosses 6 coins, he will have a 1 in 64 chance of opening the door.

Table 1.1- Information Contained in Coins

Number of Coins	information in bits	possible outcomes	Odds
0	0	2^0=1	1 in 1
1	1	2^1=2	1 in 2
2	2	2^2=2x2 =4	1 in 4
3	3	2^3=2x2x2 =8	1 in 8
4	4	2^4=2x2x2x2 =16	1 in 16
5	5	2^5=2x2x2x2x2=32	1 in 32
6	**6**	**2^6= 2x2x2x2x2x2=64**	**1 in 64**
7	7	2^7 =2x2x2x2x2x2x2=128	1 in 128
8	8	2^8=2x2x2x2x2x2x2x2 =256	1 in 256
13	13	2^{13} = 2x2x2x2.....x2= 8192	1 in 8192

Notice that there is a definite relationship between the number of bits in the door's combination and the odds that the scientist will open the door. Probability theory and information theory are closely related.

Mathematical Definition of Information

The following equation defines information:

Equation 1

$$2^{(\text{information})} = \frac{\text{Total Possible Outcomes}}{\text{Observed Outcome (s)}}$$

Example 1: a scientist is told to toss a coin 3 times and remember the results. How much information does he acquire when he observes the result? Answer: The total number of possible outcomes is 2x2x2 = 8, and only 1 outcome will be observed. So $2^{(\text{information})} = 8/1$. Because $2^3 = 2x2x2 = 8$, the scientist acquires 3 bits of information.

Example 2: Suppose the result in example 1 is H-H-T, but the scientist is unsure about the outcome because he cannot remember whether the first coin landed head or tails. How much information has he acquired? Answer: there are still 8 possible outcomes. The scientist observed either H-H-T or T-H-T, but he is not sure which. So both must be counted as observed outcomes. So $2^{(\text{information})} = 8/2 = 4$. Because $2^2 = 2x2 = 4$, the scientist acquires 2 bits of information.

Note that the odds of an event like winning the lottery are often expressed as one in some number, like a million. The one is the observed outcome, and the million represents the total number of possible outcomes. So the information acquired by knowing who wins this lottery is easy to calculate: $2^{(\text{information})} =$ (1 million outcomes / 1 outcome). Because $2^{20} = 1{,}048{,}576$, approximately 20 bits of information are acquired when the outcome of this lottery is observed.

The equation for information can be solved explicitly for information with the use of logarithms. Since many calculators have a log function, the next equation is often easier to use, but less intuitive than the first. See appendix three for a review of logarithms.

Equation 2

$$\text{information} = 3.32\text{x}\log\left[\frac{\text{Total Possible Outcomes}}{\text{Observed Outcome (s)}}\right]$$

Trapped Scientist Who Cannot See the Results

Suppose that the scientist in figure 1.1 cannot see the results of the coin toss because a screen is placed between him and the coins (figure 1.2).

Figure 1.2: Trapped Scientist Who Cannot See the Results

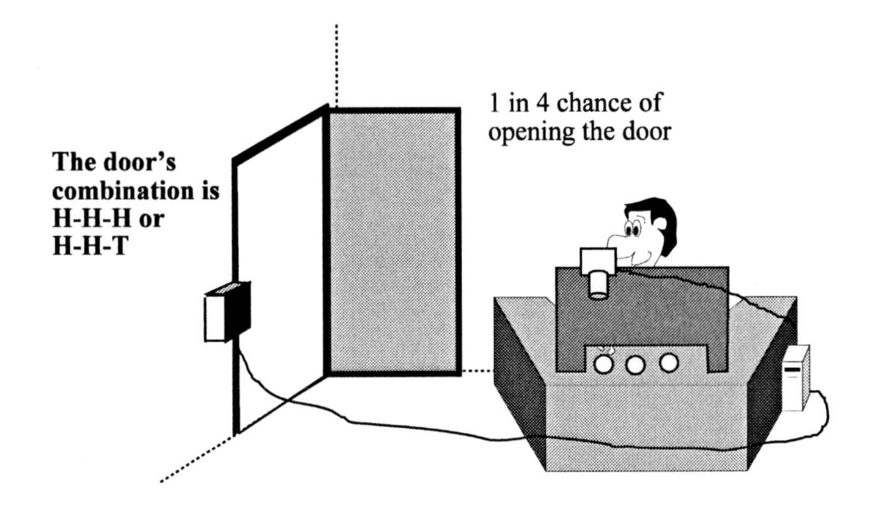

In figure 1.2, the camera monitors the results of the coin toss and sends the results to the computer. Depending on the results of the coin toss, the computer is programed to do four things: open the door, close the door, beep once, and beep twice.

Coin 1	Coin 2	Coin 3	
H	H	H	--------->opens the door
H	H	T	--------->opens the door
H	T	H	--------->closes the door
H	T	T	--------->closes the door
T	H	H	--------->computer beeps twice
T	H	T	--------->computer beeps twice
T	T	H	--------->computer beeps once
T	T	T	--------->computer beeps once

The scientist cannot observe the results of the coins. He can only observe what the door and computer do after he tosses all three coins. The door opens for 2 of the 8 possible results. If the door is already open, and the camera observes H-H-H or H-H-T then the door will stay open. Two results will close the door if it is open, and have no effect if the door is already shut. Two results will cause the computer to beep once, and two results will cause the computer to beep twice. Does the scientist still acquire 3 bits of information when he tosses three coins?

Case 1: the door opens or stays open.
$2^{(\text{information})}$ = (8 possible outcomes/2 outcomes that cause this result) =4.
Since 2^2 =4, 2 bits of information are acquired.
Case 2: the door closes or stays closed.
$2^{(\text{information})}$ = (8 possible outcomes/2 outcomes that cause this result) =4.
So in the case, 2 bits of information are acquired.
Case 3: the computer beeps once, 2 bits of information are acquired.
Case 4: the computer beeps twice, 2 bits of information are acquired.

The average amount of information acquired each time the scientist tosses all 3 coins is now 2 bits. He is using 3 coins or 3 bits to transmit 2 bits of information. He must do this because the code that translates the result of the coin toss into what the door and computer do is not the optimal code. The optimal code should only require 2 coins to transmit 2 bits. One possible optimal code is as follows:

Coin 1	Coin 2	
H	H	--------->opens the door
H	T	--------->closes the door
T	H	--------->computer beeps once
T	T	--------->computer beeps twice

The average uncertainty per symbol (or coin in this example) is called the Shannon entropy. Shannon entropy* measures on average how much each observed symbol or coin decreases uncertainty. Because information corresponds to a reduction in uncertainty, Shannon entropy is also a measure of information. When 3 coins are used to transmit 2 bits (non-optimal code), the Shannon entropy is 2/3 of a bit per coin. With the optimal code, the Shannon entropy becomes 1 bit per coin. The total information transmitted in both cases is the same because 3 coins x 2/3 bit per coin = 2 coins x 1 bit per coin = 2 bits.

*Shannon entropy should not be confused with the term entropy as it is used in chemistry and physics. Shannon entropy does not depend on temperature. Therefore, it is not the same as thermodynamic entropy.[1] Shannon entropy is a more general term that can be used to reflect the uncertainty of any system. Thermodynamic entropy is confined to physical systems.

References:

1) Brillouin, Science and Information Theory, 1956.
2) Reza, An Introduction to Information, 1961.
3) Pierce, An Introduction to Information Theory, Symbols, Signals and Noise, 1961.
4) Shannon, A Mathematical Theory of Communications, Bell Labs,1948.

Chapter 2: The Evolution of Molecular Knowledge

This chapter will introduce several examples to show that knowledge can be created by living organisms. This knowledge is created by chance in steps. The size of the step in knowledge determines whether or not chance will find an appropriate solution (figure 1 on page 6).

The Trapped Scientist

A scientist is placed in a locked room with a computer and a combination lock on the door. He is told that the door will open when he types the correct combination of words into the computer and pushes enter. He is given a basket containing 20 wooden blocks (Figure 2.1). Each block has a single word written on it. The words are as follows: *cat, drink, bike, book, apple, run, man, soon, dog, coconut, zoo, fun, radio, sun, walk, milk, water, pear, plant, computer.* The scientist is instructed to shake the basket and then select a block without looking at it. He is to read the word, enter it in the computer, and place the block back in the basket. He is further instructed to repeat this procedure until he has entered four words into the computer. He is to press enter to see if the door opens.

The combination to the door is cat-apple- *- run. The asterisk has a special meaning. At this position, any word is acceptable. There are 160,000 possible combinations, and 20 will open the door. It is very unlikely that the scientist will select the correct one on the first try. Each time he enters 4 words on the computer and pushes enter, the door has a 1 in 8000 chance of opening (1 in 20 for first word, 1 in 20 for the second, 1 in 1 for the third, and 1 in 20 for the fourth means that the odds are 1 in 20 x 20 x 1 x 20 or 1 in 8,000).

Figure 2.1: Trapped Scientist with 4 Word Combination

After a few thousand unsuccessful tries, the scientist draws cat-apple-dog-run. When he presses enter, the door opens (figure 2.1). This combination of words contains useful information, the knowledge that is needed to open the door. This knowledge was found by chance. Furthermore, this knowledge confers an advantage in that it allows the scientist to leave the room. In all of these examples, a door opening will represent a step in knowledge that confers an advantage. This book will call such a step an infon. An infon can be defined in terms of either knowledge or information.

Definition: an infon is a step in molecular knowledge.

Definition: an infon is the minimum step in information that confers a selective advantage.

Now consider the same example with a much longer combination. The combination is now drink-computer-cat-cat-bike-book-book-run-man-sun-dog-fun. The scientist is instructed to repeat the procedure of drawing words from the basket until he enters 12 words.

The number of possible combinations is now 4096 trillion. The scientist has a 1 in 4096 trillion chance of opening the door every time he presses enter. If he draws words for the rest of his life (even if he lives for a 100 million years), he will probably never open the door. The odds of him finding the correct combination are now just too remote. The knowledge needed to open the door can no longer be found by chance because the infon contains too much knowledge (figure 2.2).

Figure 2.2: Trapped Scientist with 12 Word Combination

In figure 2.2, the information per word is given by equation 2 on page 25. Information = 3.32 x log(20/1) = 4.32 bits per word. Since fractions are easier to use, 4.32 is approximately $4^1/_3$ bits; therefore, the door with a 12 word combination requires 52 bits of information to open (12 words x $4^1/_3$ bits per word). The door with the 4 word combination only needs 3 words to be correct; therefore, its combination contains 13 bits of knowledge (3 x $4^1/_3$). Figure 2.3 shows how the number of bits influences chance. Each combination is represented by a wall, and each bit adds 6 inches to the wall. If chance is represented by the scientist, then he can climb over the small wall (6.5 feet high), but he cannot climb over the 26 foot wall.

Figure 2.3: Knowledge Represented by a Wall to Climb

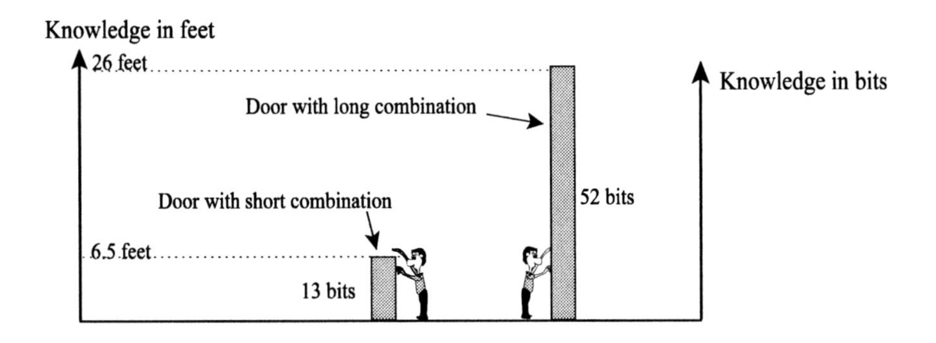

Information, Knowledge, Complexity and Order

Complexity and order are often confused as the same thing, but they are antonyms. Likewise, information and complexity are often considered the same, but they are not. Using the trapped scientist, these terms will now be defined.

Order is the easiest to understand. If the scientist draws the word cat 12 times in a row from the basket and enters this into the computer, then the message he enters is ordered. Ordered messages (or combinations in this example) contain patterns that allow them to be simplified. The combination cat-dog-cat-dog-cat-dog is ordered.

Complex messages are messages that are not ordered. The combination cat-drink-bike-book-apple-run-man-soon-dog-coconut-zoo-fun is complex because it is not ordered.

Information is any change in uncertainty. Once the scientist observes the result of any draw, he acquires information. The results do not influence information because each word reduces his uncertainty by the same amount. Only the combination that opens the door conveys knowledge. In figure 2.2, this combination is also complex, but it does not have to be.

The term specified complexity[1] is often used in intelligent design literature. This terminology is misleading because biological messages do not have to be complex. They can also be ordered.

If the scientist predicts that he will enter cat-dog-man-fun-plant before drawing any blocks, then he has specified a complex message, but unless he specifies the combination of the door, this message will not be useful, and it will not confer a selective advantage. Specified complexity is only relevant to evolution if the word *complexity* is interpreted to mean "a message that confers a selective advantage." The word *knowledge* has this meaning already built into its definition. Knowledge is always specified and useful, and in living things for knowledge to be useful, it must confer a selective advantage.

A Scientist Locked in a Room with Multiple Doors

The scientist is now locked in a room with four doors. He is told that if he opens any door, the computer will let him know the combination for that door by locking the words responsible for opening the door. The combinations are as follows:

Inner door	drink-computer-cat-*- *-*- *- *- *-*- *.
Second door	drink-computer-cat-cat-bike-book-*- *-*-*-*-*.
Third door	drink-computer-cat-cat-bike-book-book-run-man-*-*-*.
Last door	drink-computer-cat-cat-bike-book-book-run-man-sun-dog-fun.

After a few thousand tries, the scientist enters drink-computer-cat-cat-bike-man-sun-dog-dog-cat-drink-run (figure 2.4). The first door opens, and the computer locks the combination of this door. The scientist can no longer change the first three words. So instead of picking 12 words from the basket, he now picks 9. He enters these into the computer and presses enter. After a few thousand tries, he enters cat-bike-book-fun-run-man-fun-dog-dog. The second door opens and the computer locks the first 6 words, so that they cannot be changed. The scientist continues now drawing 6 words. When he enters book-run- man-dog-cat-cat, the third door opens. The computer locks the first 9 letters, and the scientist continues. When he enters sun-dog-fun the last door opens (figure 2.5).

Figure 2.4: Trapped Scientist with 4 Doors and Short Combinations

Figure 2.5: Trapped Scientist with 4 Doors Open

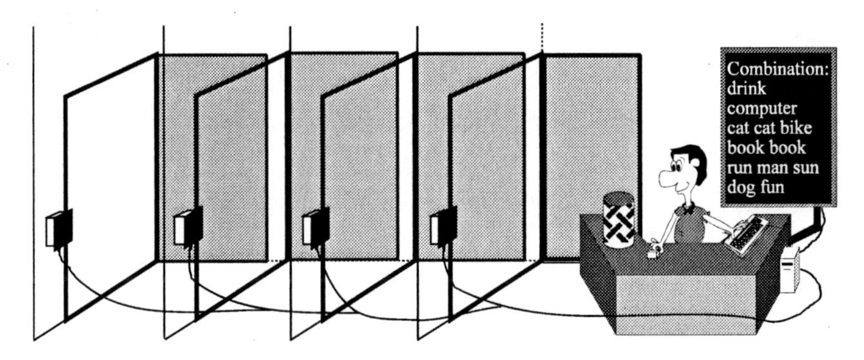

Notice that the knowledge required to open the last door in figure 2.5 is identical to that required in figure 2.2. The scientist finds the knowledge required in this example, and he fails in the previous one. Why? Each door represents a step in knowledge. All steps in figure 2.4 and 2.5 are small (3 words). Furthermore, because the computer preserves the combinations that open doors, only combinations that are close to the desired combination are preserved.

The same amount of knowledge is needed to open the last door in this example, but the correct combination is found because the steps in knowledge needed to find it are kept small by using 4 doors instead of one. Now suppose the combinations are as follows:

Inner door — drink-computer-cat-cat-*-book-*- run- man-*-dog-fun.

Second door — drink- computer- cat- cat- bike- book-*-run-man-*- dog-fun.

Third door — drink-computer-cat-cat-bike-book-book-run-man-*- dog-fun.

Last door — drink-computer-cat-cat-bike-book-book-run-man-sun- dog-fun.

The odds of opening the first door are now 1 in 512 billion. The scientist never opens the first door (Figure 2.6).

These examples show that the number of new words needed to open a door determines whether or not chance will open the door. Each combination represents a step in knowledge. If the steps are small, (three new words or less), then the combinations are easily found by drawing the words from the basket. If the steps are large, (nine new words or more), then chance can no longer reliably find the combination.

Figure 2.6: Trapped Scientist with 4 Doors but One Large Step

When Darwin introduced the theory of evolution, he did not consider information, but he did mandate that the changes must be slight and continuous. So he did at least understand the nature of the problem. Darwin's premise is simple. Small changes that provide an advantage are preserved by nature. Over millions of years, these changes are cumulative and thus lead to very large changes. His theory works if and only if the steps in knowledge are small.

The scientist in the locked room with 4 doors models natural selection. When the scientist opens a door, the computer preserves the combination that opened the door. That is once a small amount of knowledge is created by chance, the computer preserves it. This preservation is fully analogous to natural selection in biological systems. With the help of natural selection, the scientist can easily find the combination to a hundred doors or more (as long as the combinations are three words or less). But if the first door's combination is large, 9 words or more as in figure 2.6, then chance is no longer effective. All doors remain closed.

Natural selection is not effective when the steps are large because chance never finds the correct combination, and there is no knowledge for natural selection to preserve. So the size of the first step is critical. It completely determines whether or not the scientist can find the correct combination.

Just like before, each door can be represented by a wall (figure 2.7). Since there are 4 doors, there are 4 walls. The walls are now pushed against each other so that they form a series of steps. If all of the steps are small then the scientist can easily climb to the top - even if there are a thousand steps, but just one large step can present a serious problem. The 19.5 foot wall is very difficult for the scientist to climb. For the scientist to get over this wall, he will need some help.

Figure 2.7: Steps in Knowledge: Doors Represented as Walls

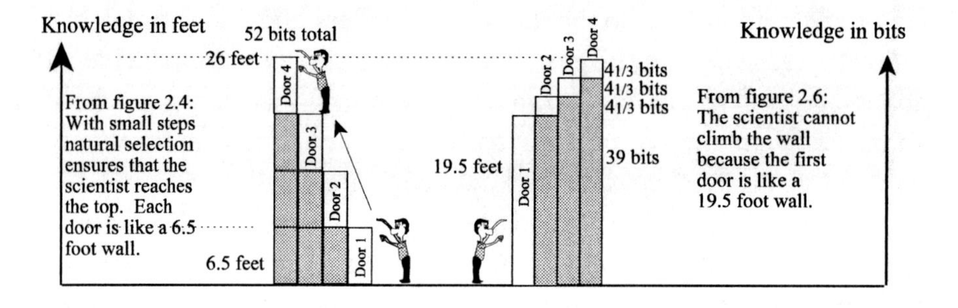

Important Definitions:

Darwinian evolution - the steps in knowledge are small so chance creates useful information and this knowledge is preserved by natural selection. Darwinian evolution happens just like Charles Darwin theorized.

Evolution by design - the steps in knowledge are very large; as a result, chance never creates any new useful information. Natural selection is irrelevant because no knowledge is created for it to preserve and optimize.

The goal of this book is to analyze the steps in knowledge required for evolution. The results will show that many steps are similar to a locked door with a fifty word combination. Others are similar to a locked door with a one word combination. The small steps are navigated just like Darwin theorized. Small steps do not present a barrier to chance. On the other hand, the large steps are a significant barrier. How evolution crossed this large barrier is not known.

Probability and Information

The information found in a molecule of DNA or in a protein can only be associated with a probability for evolution when natural selection is excluded. Because of natural selection, it is not possible to relate the information found in the combination of the last door to a probability of evolution. The trapped scientist examples do an excellent job of explaining why this is true (compare figure 2.2 to figure 2.5). The last door in both figures has the same combination, yet one is opened easily and the other remains closed.

A probability can be associated with each individual door opening because by definition natural selection is not active before the molecular knowledge exists. Obviously, the door with the largest step in knowledge will determine whether or not the scientist escapes. This is always the door with the most unknown words (it is seldom the door with the longest combination). In most cases, the odds associated with the evolution of knowledge will depend entirely on a single door, and this is usually the first door because this is the door that creates the initial knowledge (refer to figure 1 on page 6 and figure 2.7). All doors after the initial door only need to optimize the existing knowledge. Thus, these steps are generally much smaller.

Information in Biological Systems:

Life does not use words like *cat*, *dog* and *computer*. The words that life uses are chemicals called amino acids. Amino acids are the building blocks of proteins. Proteins implement the molecular knowledge that enables life to live. DNA merely contains the information needed to build proteins. Proteins do all of the work.

The trapped scientist example uses 20 words because there are 20 amino acids used by life to build proteins. Each word corresponds to an amino acid. So if the 20 words in these examples are replaced with the names of the 20 amino acids, then the examples will more accurately model the evolution of a protein. In this example, the combination of the doors is composed of the following words: serine, arginine, proline, leucine, valine, isoleucine, alanine, glycine, cysteine, lysine, tryptophan, tyrosine, methionine, glutamate, aspartate, asparagine, glutamine, hystidine, threonine, and phenylalanine. These are the names of the 20 amino acids that are found in proteins. Each amino acid is represented with a three letter abbreviation: ser, arg, pro, leu, val, ile, ala, gly, cys, lys, trp, tyr, met, glu, asp, asn, gln, his, thr, and phe. The blocks in the basket now have these abbreviations painted on them. Nothing else has changed.

Suppose for the protein under consideration to have any function, 27 of the 30 amino acids must be correct. This corresponds to a combination for the first door as follows: met-phe-his-*-lys-pro-ser-*-val-ala-lys-trp-asp-asp-phe-met-gln-his-lys-cys-his-thr-*-gln-lys-pro-pro-ala-ala-gln. Once this protein is created by chance, natural selection takes over, preserving and optimizing the sequence. The asterisks are eventually changed to the correct amino acids as shown on the last door in figure 2.8. The process of asterisk replacement optimizes the protein.

Since the protein exists, the scientist should be able to break the combination. So does he ever find the combination to the first door? Given 50 billion years with 100 million tries a year, the probability of the scientist opening the first door is only 1 in 27,000 trillion. He does not open the door.

Figure 2.8: Using Amino Acids for Words

Reference:

1) Dembski, Intelligent Design, InterVarsity Press, 1999.

Chapter 3: Information Storage and Transfer in Life

The trapped scientist examples are great for conceptual purposes, but they do not accurately model how information in life changes because they do not take into account the fact that amino acid changes are caused by changes in DNA. This chapter will explore how DNA stores information, and how this information is used to build proteins. It will also explore how mutations change this information.

The language that life uses to store and transmit information is similar to human languages, but the rules of grammar and the vocabulary are much simpler. Only 20 words are used by life, so the vocabulary is very limited. Punctuation is limited to capitalization and periods. Every sentence must start with the same word.

This chapter will start with a simple system based on coin tosses and show how coin tosses can be used to store and transmit information. This simple example will then be improved by using a four sided coin. The trapped scientist will then be used with a new set of rules to show how information changes in life.

Transcription and Translation

Suppose that the phrase uno, dos, tres is found in a Spanish book. The process of transcription copies a page from this book onto a piece of paper. Translation then requires a person with a Spanish to English dictionary to copy the phrase onto another piece of paper in English (figure 3.1). Life uses both transcription and translation. The message in the DNA corresponds to the written words in a Spanish book (uno, dos, tres). This message is copied through transcription to create an intermediate message called RNA. RNA is analogous to the piece of paper with the phrase uno, dos, tres. The final piece of paper with the phrase one, two, three is analogous to a protein.

Figure 3.1: Transcription and Translation

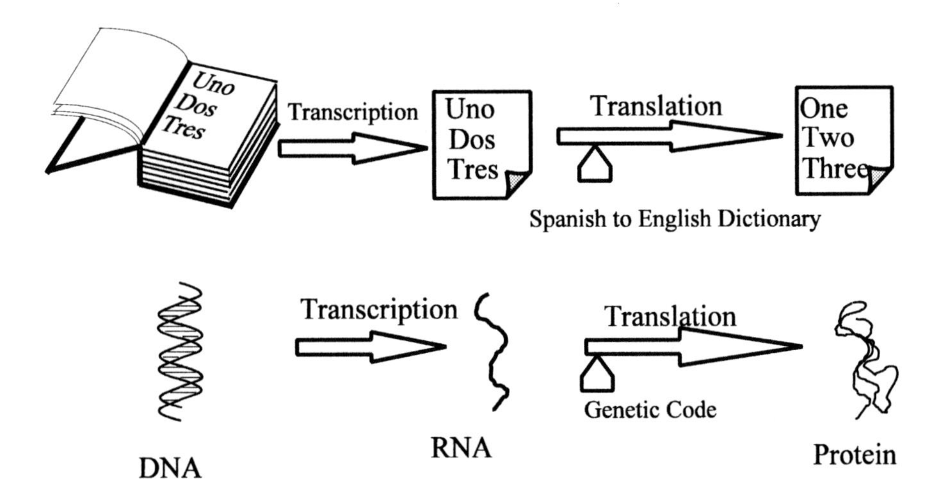

Information Using Coins

Coins have two possible outcomes when tossed, heads and tails. So to store and transmit more than two possible messages, the results of tossing a coin must be grouped. If coins are tossed and read three at a time, then eight messages can be assigned to 3 coins.

The code is important because it assigns the messages. The code performs the function of the Spanish to English dictionary in figure 3.1. Consider the following code, and how it can be used to create a message (figure 3.2).

Coin 1	Coin2	Coin 3		Message
T	T	T		The
T	T	H		Cat
T	H	T		Dog
T	H	H	–coin code---->	Is
H	T	T		Big
H	T	H		Small
H	H	T		Tall
H	H	H		Short

Figure 3.2: How Coins Can be Transcribed and Translated

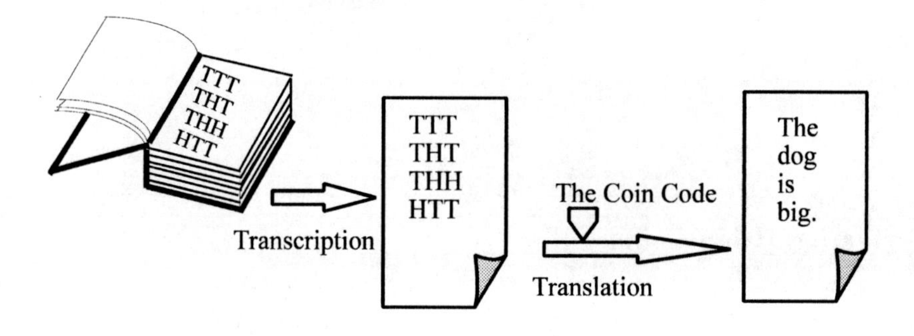

A Four Sided Coin

Life uses the equivalent of a four sided coin. To imagine this coin, visualize a playing die. Instead of numbers on each side, each side has a single letter. The letters are A,T,C and G. Since dice have 6 sides, this die has to be modified. This is easy to do. Imagine a long pin inserted into one side and out the other. This pin only allows the die to land on the 4 sides with letters. See figure 3.3.

Figure 3.3: Four Sided Die (Coin) with the Letters A, T, G, and C

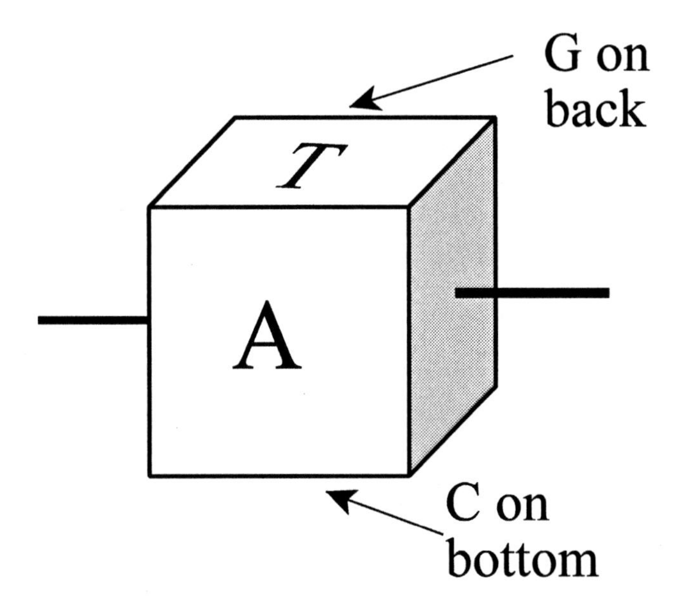

If this altered die is used as money, then it is just as good as any other 4 sided coin. But now instead of each coin having two possible outcomes, head or tails, each coin has 4 possible outcomes, A, T, G and C. Thus, each coin now contains 2 bits of information because $2^2 = 4$ (equation 1, p24).

If these special coins are grouped three at a time, how many possible outcomes are there? Each coin contains 2 bits of information. Thus, 3 coins must contain 6 bits (2+2+2=6). There are now 64 possible outcomes because $2^6 = 64$. This means that life could potentially have at its disposal 64 words, but life only needs 20 words. So some outcomes are assigned the same word.

The Four Sided Coin Code

The four sided coins are always tossed and read 3 at a time. The following table assigns a word to each possible outcome. Since only 20 words are used, some outcomes are assigned to the same word. The period in the parenthesis is not a word, but indicates the end of a sentence. All sentences must start with the word, *the*, and this is why it is capitalized. The table is like a Spanish to English dictionary. It assigns a message to an outcome.

Table 3.1: The Coin Code

Outcome	Message
CCA, CCG, CCT, CCC	dog
CGA, CGG, CGT, CGC	cat
CAA, CAG, CAT, CAC	bird
GAA, GAG, GAT, AAC, GAC, AAT	animal
TAA, TAG, TAT	is
GGA, GGG, GGT, GGC	runs
ACC	sleeps
AAA, AAG	under
TAC	The (Capitalize)
AGA, AGG, AGT,AGC, TCA, TCG	lazy
TGA, TGG, TGT, TGC	quick
GAT, GTG	big
GTT, GTC	tall
CTT, CTC	small
CTA, CTG	tiny
TTA, TTG	tree
TTT, TTC	buried
ACA, ACG	home
ATA, ATG	white
GCA, GCG, GCT, GCC, TCT, TCC	black
ATC, ACT, ATT	(Period)

Using this table, it is possible to construct many sentences. Consider the following outcomes and the associated messages. These messages are constructed by tossing the special coin 15 times and then grouping the results into triplets. Each triplet is called a codon.

TAC-CCA-TAA-AGA-ATC => The dog is lazy.
TAC-CGA-TAA-GAT-ATC => The cat is big.
TAC-GAA-TAA-CTG-ATC => The animal is tiny.

Many more nonsense messages are possible.

TAC-ATA-GCA-CTA-ATC => The white black tiny.
TAC-ATA-TAA-CTA-ATC => The white is tiny.

Also notice that the reading frame is very important. Consider the following message:

TAC-CCA-TAA-AGA-ATC => The dog is lazy.

If the first *T* is deleted, then the message becomes ACC-CAT-AAA-GAA-TC => sleeps bird under animal. This sentence is nonsense. To get the correct message the letters must be read three at a time, and the reading frame must be correct. Shifting the frame changes how the letters are grouped. Thus, it changes the words, and very simple change like deleting a single letter can have far reaching consequences. Because such mutations change the reading frame, they are called frame shift mutations. In contrast, point mutations only change a single base and thus preserve the reading frame.

The Information in DNA is Similar to the Four Sided Coin

Life uses a system very similar to the four sided coin, but instead of coins and dictionaries, life uses chemicals. DNA looks like a twisted ladder. The steps of the ladder are composed of four chemicals, adenine, thymine, cytosine and guanine. These four chemicals correspond to the four letters on the four sides of the four sided coin. That is adenine is A, thymine is T, cytosine is C and guanine is G. The one letter abbreviations for these chemicals will be used from now on. The side of the ladder is composed of a chemical called deoxyribose.

Notice that the steps are A-T, T-A, G-C or C-G, where the dash represents a chemical bond between the chemicals. Adenine always forms a chemical bond with thymine, and guanine always forms a bond with cytosine. A-T is called a base pair as is G-C. There are four possibilities for each step, so each step can hold the same amount of information as the four sided coin or 2 bits.

Figure 3.4: Untwisted DNA Ladder

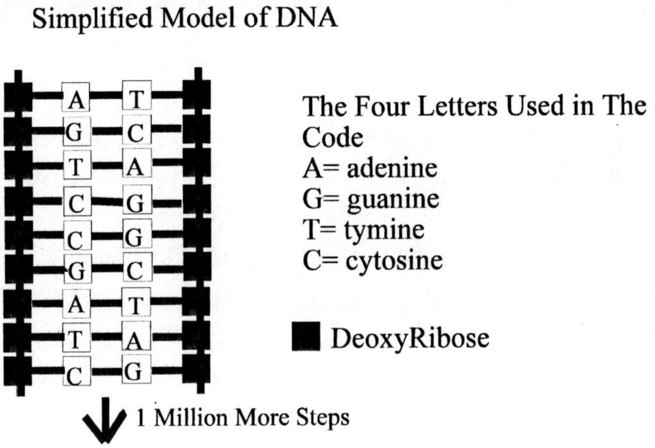

Simplified Model of DNA

A-T
G-C
T-A
C-G
C-G
G-C
A-T
T-A
C-G

1 Million More Steps

The Four Letters Used in The Code
A= adenine
G= guanine
T= tymine
C= cytosine

DeoxyRibose

Figure 3.5: Twisted DNA Ladder

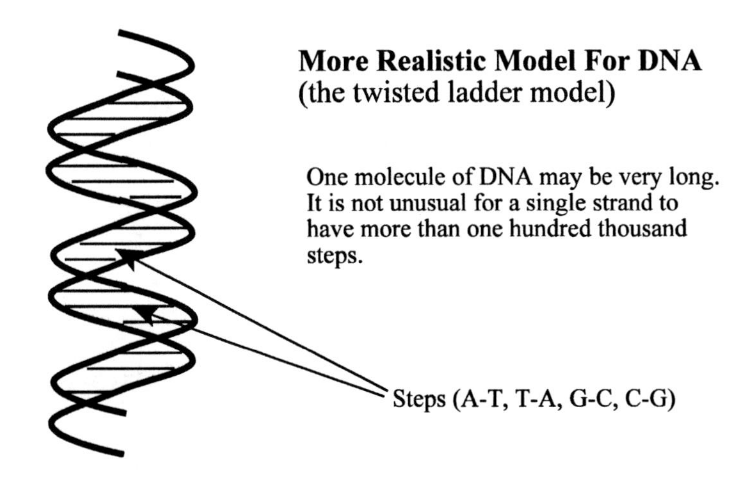

DNA Replication

In order to replicate, the DNA molecule must be untwisted, and the chemical bonds between the base pairs must be broken. This process is controlled by many different proteins. In the first step, a protein binds to the DNA targeting the site to be replicated, another protein untwists the DNA breaking the chemical bonds between the base pairs, other proteins keep the base pair bonds from reforming, and a protein called DNA polymerase replicates the untwisted sections. The two original strands serve as templates (solid lines in figure 3.6) for the growing strands (dashed lines). Figure 3.6 shows how DNA polymerase replicates each strand, creating two DNA molecules from one. Figure 3.6 is simplified in that most of the proteins involved in DNA replication are not shown.

Figure 3.6: DNA Replication

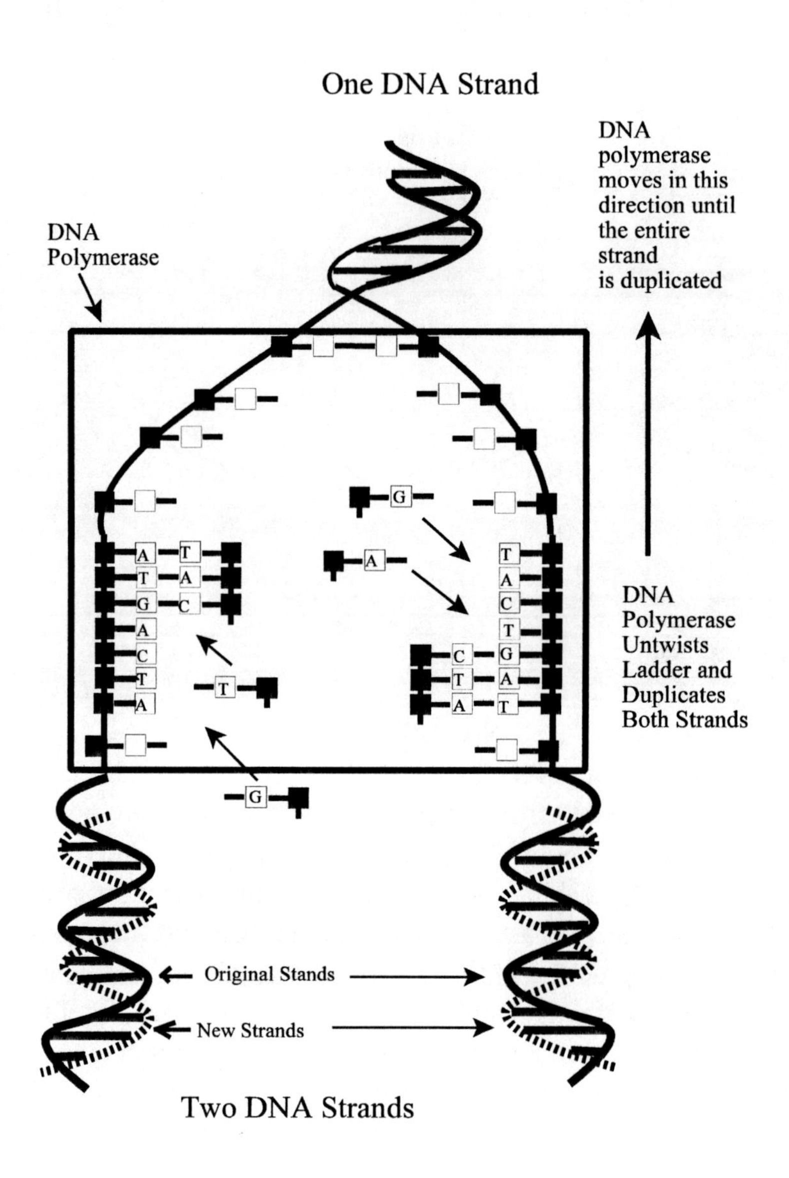

During DNA replication, sometimes the bases pair incorrectly. That is sometimes adenine pairs with guanine, and thymine pairs with cytosine. Proofreading corrects most of these errors. Thus, DNA replication is very accurate and mistakes are rare, but mistakes happen. Most mistakes just create variability, but some can create information.

The Genetic Code

The coin code in table 3.1 applies to the four sided coin. The genetic code works in the same way, except instead of words the code specifies amino acids. The coin code groups the results of the coin tosses into groups of three. The genetic code does the same, except the letters are now chemicals. A group of three bases is called a codon. Each codon specifies an amino acid. For example, with the coin code:

TAC-CCA-TAA-AGA-ATC should be translated as follows: The dog is lazy.

but with the genetic code (see table 3.2)

TAC-CCA-TAA-AGA-ATC should be translated as follows: methionine-glycine- isoleucine-serine.

So the coin code specified words, and the genetic code specifies amino acids. Table 3.2 lists the genetic code, and figure 3.7 shows the process of transcription in which DNA is use to create RNA. Notice that in figure 3.7 the coding DNA strand creates a complementary RNA strand (G is replaced with C, C is replaced with G, T is replaced with A, and A is replaced with a new chemical unique to RNA - uracil or U for short).

Table 3.2: The Genetic Code

DNA Codon (Coding Strand)	Corresponding RNA Codons (Transcription)	Amino Acid (Translation)
CCA, CCG, CCT, CCC	GGU, GGC, GGA, GGG	glycine
CGA, CGG, CGT, CGC	GCU, GCC, GCA, GCG	alanine
CAA, CAG, CAT, CAC	GUU, GUC, GUA, GUG	valine
GAA, GAG, GAT, AAC, GAC, AAT	CUU, CUC, CUA, UUG, CUG, UUA	leucine
TAA, TAG, TAT	AUU, AUC, AUA	isoleucine
GGA, GGG, GGT, GGC	CCU, CCC, CCA, CCG	proline
ACC	UGG	tryptophan
AAA, AAG	UUU, UUC	phenylalanine
TAC	AUG	methionine
AGA, AGG, AGT,AGC, TCA, TCG	UCU, UCC, UCA, UCG, AGU, AGC	serine
TGA, TGG, TGT, TGC	ACU, ACC, ACA, ACG	threonine
GAT, GTG	CUA, CAC	histidine
GTT, GTC	CAA, CAG	glutamine
CTT, CTC	GAA, GAG	glutamate
CTA, CTG	GAU, GAC	aspartate
TTA, TTG	AAU, AAC	asparagine
TTT, TTC	AAA, AAG	lysine
ACA, ACG	UGU, UGC	cysteine
ATA, ATG	UAU, UAC	tyrosine
GCA, GCG, GCT, GCC, TCT, TCC	CGU, CGC, CGA, CGG, AGA, AGG	arginine
ATC, ACT, ATT	UAG, UGA, UAA	Stop

Transcription

Transcription copies the information in DNA into RNA as shown in figure 3.7. Note that RNA is a single strand, whereas DNA is double stranded. Also note the RNA does not contain the chemical thymine. Another chemical called uracil replaces it. Thus, an A-U bond (adenine-uracil) is formed between the DNA and RNA strand (as opposed to adenine-thymine). The enzyme RNA polymerase directs the synthesis of RNA.

Figure 3.7: Transcription

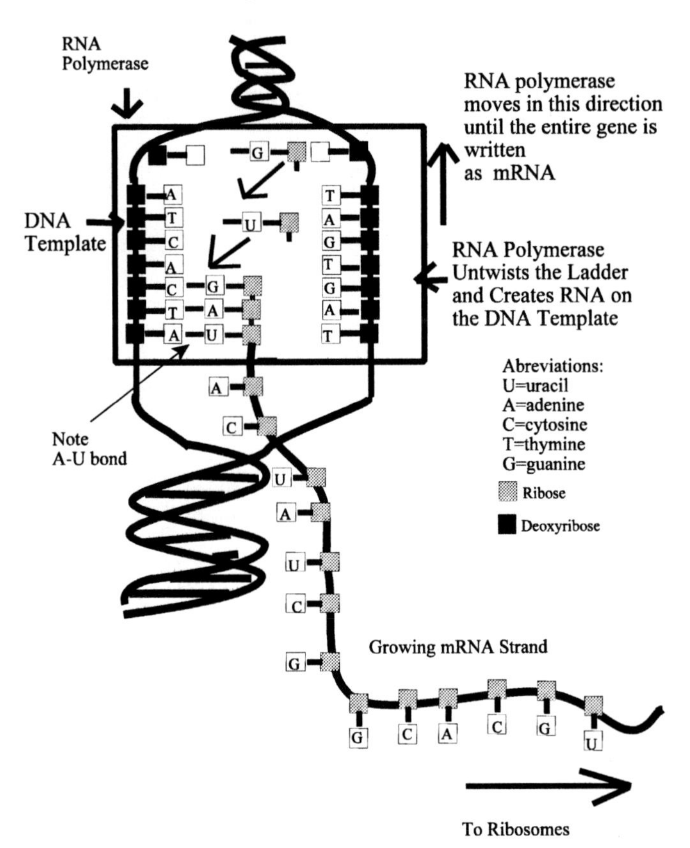

Each Codon Specifies an Amino Acid

Each codon in a gene specifies a particular amino acid. Therefore, a gene determines the chemical properties of a specific protein by specifying its amino acid sequence. A simplified example is shown in figure 3.8. Four DNA codons and the amino acids that they specify are depicted in this picture. During transcription of the hypothetical DNA sequence, the DNA is written into a corresponding messenger RNA (mRNA). During translation, mRNA is translated using the genetic code to create a new protein. The code assigns an amino acid to each codon.

The properties of amino acids are determined by their side chains. In figure 3.8, these side chains branch from the main protein chain. The side chains of amino acids give them unique chemical properties. 1) Some side chains are chemically reactive. These side chains are used by proteins to interact with other chemicals. 2) Other side chains do not like to dissolve in water. Such side chains are called hydrophobic. Hydrophobic side chains cause proteins to fold up into complex three-dimensional shapes.

Figure 3.8: Translation

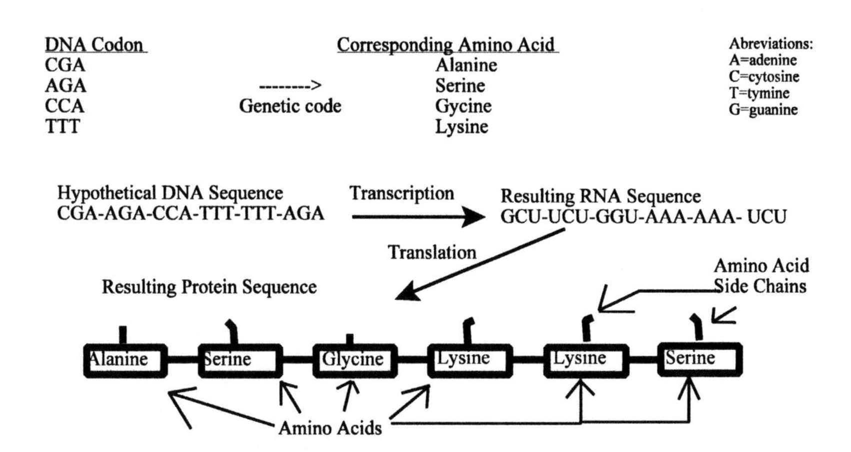

To verify understanding, all readers should make sure that they can use table 3.2 to follow how the hypothetical DNA sequence in figure 3.8 is transcribed into the resulting RNA sequence, and how this messenger RNA sequence is translated into the final amino acid chain.

Translation

Messenger RNA is translated at ribosomes. Figures 3.9-3.12 illustrate this process. A special type of RNA known at tRNA recognizes amino acids and brings them to the ribosome. The tRNA matches up its own sequence of 3 bases to the codon in the messenger RNA.

Figure 3.9: Transfer RNA Brings and Aligns Amino Acids

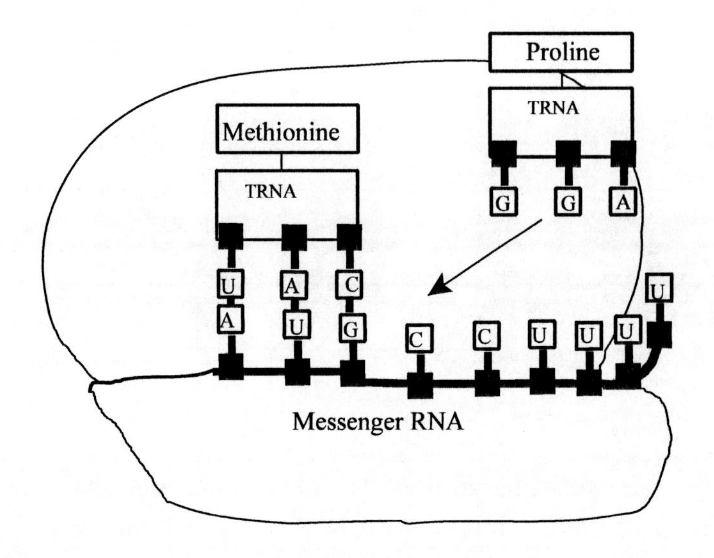

Figure 3.10: A Peptide Bond Forms between the Amino Acids

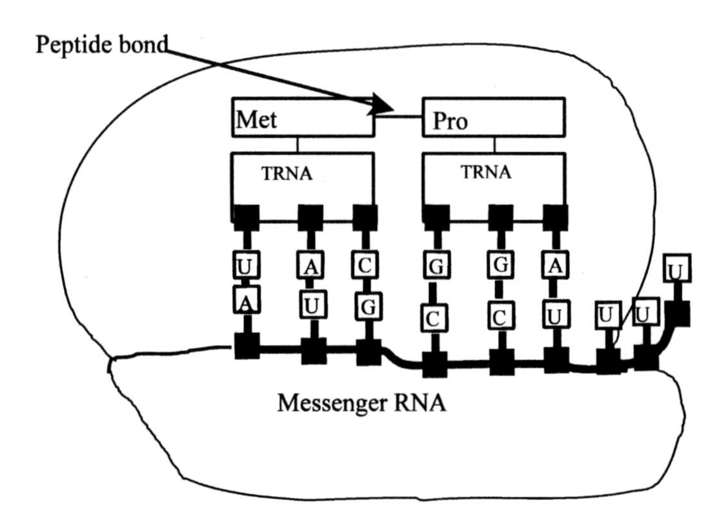

Figure 3.11: Ribosome Shifts Along Messenger RNA

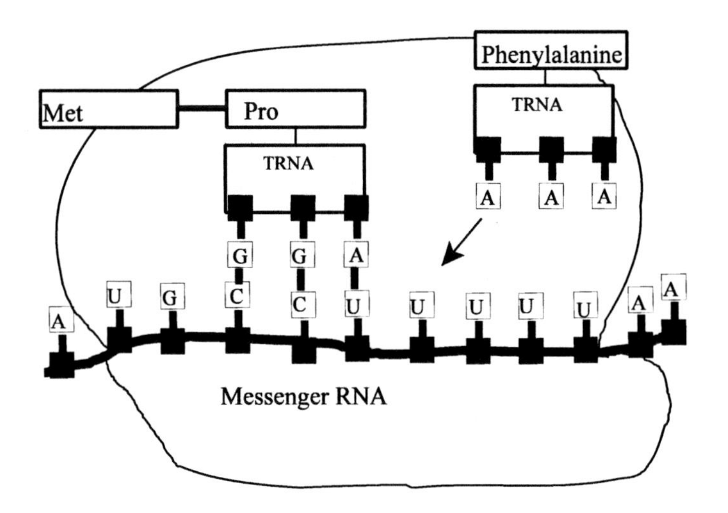

Figure 3.12: Second Peptide Bond Forms and Cycle Starts Over

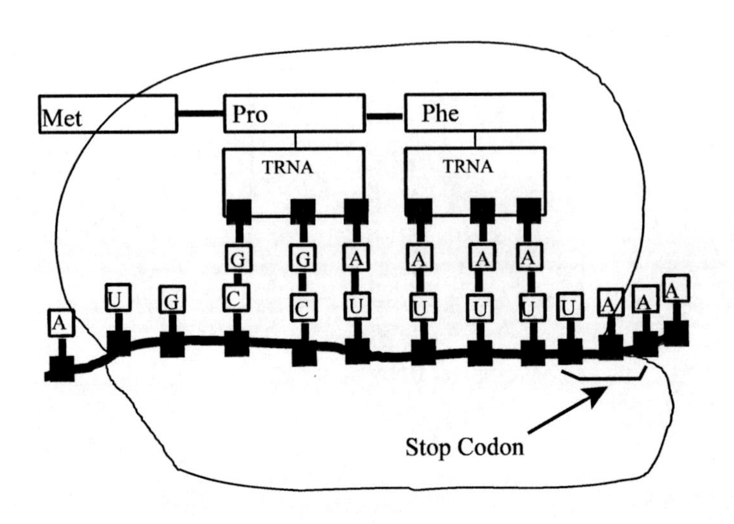

Very Simple Peptide

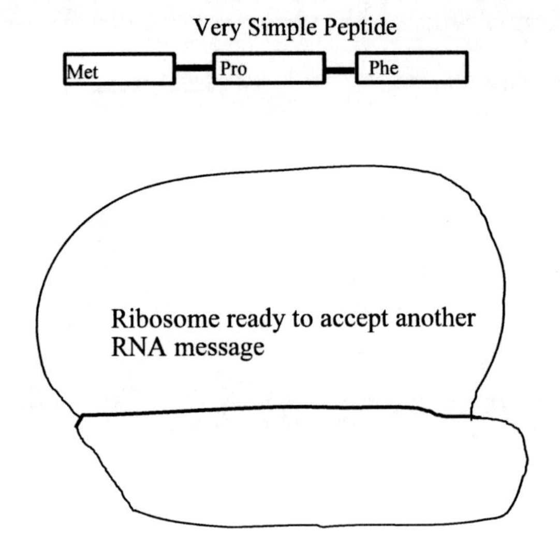

Proteins

Proteins are the most versatile chemicals found in life. Proteins are so important because they are involved in everything that life does. They implement the knowledge contained in DNA. Some of the primary functions of proteins are listed below.

A special kind of protein called an enzyme regulates chemical reactions. Chemical reactions that would take years without enzymes proceed in fractions of a second with enzymes. Enzymes make life possible. Teams of enzymes working together enable cells to synthesize all sorts of complex chemicals.

Proteins have additional functions as well. Some proteins regulate genes. Others control which chemicals can pass though the cell membrane, and still others are responsible for movement (muscles are composed of two proteins, actin and myosin). The list does not stop here. Some proteins can transport other chemicals. Hemoglobin is the blood protein that transports oxygen. Proteins also serve as signals. For example, insulin signals cells to take up sugar from the blood stream. Diabetes results when this process does not function properly.

Proteins are very versatile molecules, and it is this versatility that allows proteins to implement the knowledge stored in DNA.

Protein Folding

Proteins fold into complex 3 dimensional patterns. The amino acid sequence of a protein determines this pattern. The amino acids that do not like water tend to cluster in the protein's core where water is excluded. Amino acids that like water are typically found on the surface of the protein where they can interact with water. A random amino acid sequence will rarely fold properly into a compact 3-D shape. So the amino acid sequence is very important to ensure that a protein folds properly.

Some segments of a protein may form helixes. Others form sheets. Because of the complexity, the structure of proteins is best illustrated with cartoons. Rather than drawing the amino acids that form the helix, the cartoon model just draws the helix. The sheets are drawn as straight lines with arrows on the end. The protein bacterial rhodopsin is shown in figure 3.13. Notice the helixes. Also note that black and gray are used to represent amino acids that contain information.

Figure 3.13: Bacterial Rhodopsin

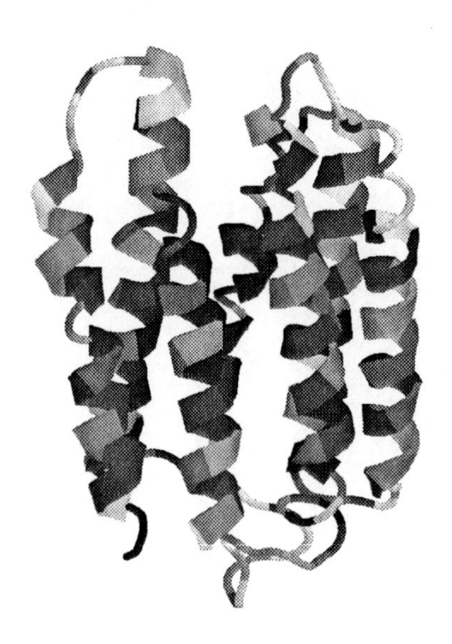

The protein that composes the eye lens in mammals is called crystallin. It tends to form sheets instead of helixes. Its cartoon representation is shown in figure 3.14. Again, black and gray represent amino acids that carry information. Figure 3.14 also shows crystallin using the space filling model for the atoms (right side). That is all of the atoms in the protein are represented by spheres. It is very hard to visualize the structure of a protein using this model. This is why the cartoons are so useful. When viewing these structures, keep in mind that they are formed by hundreds and sometimes thousands of amino acids.

Figure 3.14: Carton and Space Fill Model for Crystallin

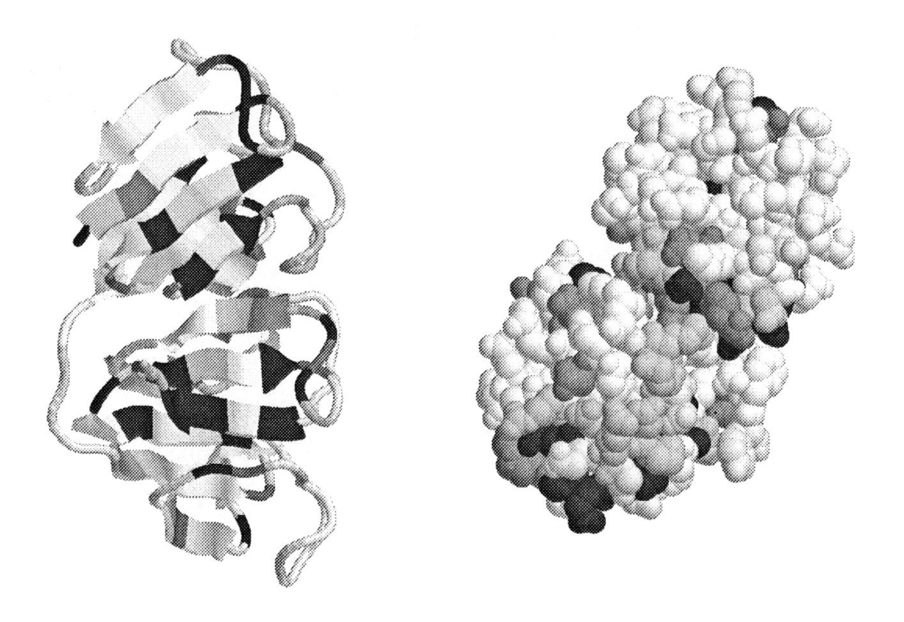

The top right picture on the back cover shows bacterial rhodopsin in color. Purple and shades of purple are regions rich in information. This image was created by applying a script generated by the CONSURF program to the protein data bank file and then viewing the results in the RASMOL program (see page 288).

Eukaryotes and Prokaryotes

Figure 3.15 depicts a eukaryotic and a prokaryotic cell. The DNA in eukaryotes is bundled together with proteins to form chromosomes. This DNA resides inside the nucleus which is separated form the rest of the cell by the nuclear membrane. In prokaryotes, there is no nucleus. The DNA is usually one large circular ring, and the cell contains much less DNA. Prokaryotes are very simple organisms such as bacteria. The total information contained in their DNA is an order of magnitude less than the information found in eukaryotes. The mitochondria in eukaryotes converts sugar into ATP - the fuel for the cell. In bacteria, proteins in the cell membrane create ATP.

Figure 3.15: Prokaryotes and Eukaryotes

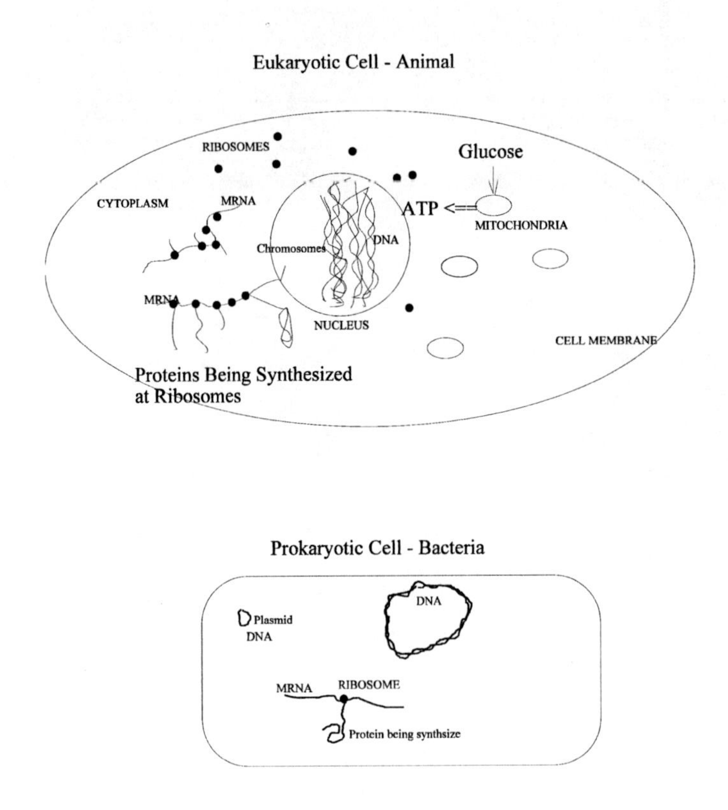

The Accurate Trapped Scientist

The previous trapped scientist models are oversimplified because they model changes to amino acid sequences without considering DNA. Since mutations alter DNA, the trapped scientist really should change DNA.

When considering DNA, another factor will surface. Six codons specify the amino acid, arginine. Only one codon specifies methionine. Three codons specify isoleucine. From table 3.2, each amino acid may have 1,2,3,4 or 6 corresponding codons. If the bases A, T, G and C are changed at random, the probability of creating a codon that specifies arginine is much higher than the probability of creating a codon that calls for methionine. Information theory must take this effect into account when computing the information content of a protein. For simplicity, only point mutations that change A, T, C or G will be considered (no insertions or deletions allowed), and all mutations will be considered random.

Figure 3.16: Trapped Scientist Using Genetic Code

The doors combination is met-ala-val his-cys-lys

1 in 537 million chance of opening each door

Combination:
TAC-CGA
CAA-GAT
ACA-TTT

The scientist now requires two baskets. In one he has four blocks labeled A, G, C and T. In the other, he has 18 blocks numbered 1 through 18. He is told to pull a lettered block and enter it into the computer. He is to put the block back and repeat this procedure. After he enters 18 letters, he is to press enter to see if the door opens. If the door does not open, he is to draw one lettered block and one numbered block. He is to use the number to find a corresponding letter on the computer screen (the letters on the computer screen are numbered sequentially 1 through 18). After he finds the letter corresponding to the number, he is to replace this letter with the new letter. For example, in figure 3.16, the last letter on the computer screen is *T*. If the scientist draws the number 18 and the letter A, then he should change the letter *T* to an *A*. He should repeat this procedure until he opens the door.

The probability that he will open the door is now a little more complex to calculate. The door will open if the scientist enters any sequence of letters that specifies the combination of the door, methionine-alanine- valine-histidine-cysteine -lysine. Using table 3.2 for reference, there are 64 possible codons that can be typed into the computer. One of these codons specifies methionine. Four of these codons specify alanine and valine. Two codons specify histidine, cysteine and lysine.

Methionine has a 1 in 64 chance of arising by chance. The knowledge it contributes is $2^{(\text{information})} = 64$. Since $2^6 = 64$, methionine contributes 6 bits. Alanine and valine each have a 1 in 16 chance of arising by chance. The knowledge they contribute is thus $2^{(\text{information})} = 16$. Because $2^4 = 16$, each contributes 4 bits. Histidine, cysteine, and lysine each have a 1 in 32 chance of arising by chance. Because $2^5 = 32$, each contributes 5 bits. Thus, the total knowledge required to open the door is $6 + 4 + 4 + 5 + 5 + 5 = 29$ bits. The odds that the scientist will open the door on the first try are 1 in 2^{29} or 1 in 537 million.

Chapter 4: Information and Knowledge in the Protein Insulin

This chapter will calculate the information and molecular knowledge in a real protein. The techniques discussed in this chapter to calculate knowledge are somewhat arbitrary because they rely on both math and human insight. Furthermore, the techniques used here to calculate information differ slightly from those used by other authors. For justification, interested readers are referred to appendix 2.

For this chapter, a small protein is desirable. Insulin meets this criteria. Insulin is a special kind of protein known as a hormone. When insulin is released into the blood stream, it signals cells to absorb sugar. The actual hormone consists of two short chains, A and B. This chapter will calculate the information and knowledge in both chains. The A chain contains 21 amino acids and the B chain contains 30. The B chain will be considered first.

The most common sequence for chain B in mammals is as follows:

phe-val-asn-gln-his-leu-cys-gly-ser-his-leu-val-asp-ala-leu-
tyr-leu-val-cys-gly-glu-arg-gly-phe-phe-tyr-thr-pro-lys-ala

At several positions in this chain more than one amino acid is allowed. The number of allowed amino acids at each position is determined by comparing the insulin found in man to that in pigs, cats, dogs, fish, frogs and snakes. The next section will illustrate this technique.

Determining Allowed Amino Acids

Because insulin exists, the doors are open in figure 4.1, and the scientist has left. In all three cases, the door's combination is 7 amino acids long. The combination that opened each door is shown on the screen. The combinations are very similar, but there are differences.

Figure 4.1: A Section of Insulin in Cat, Snake and Fish

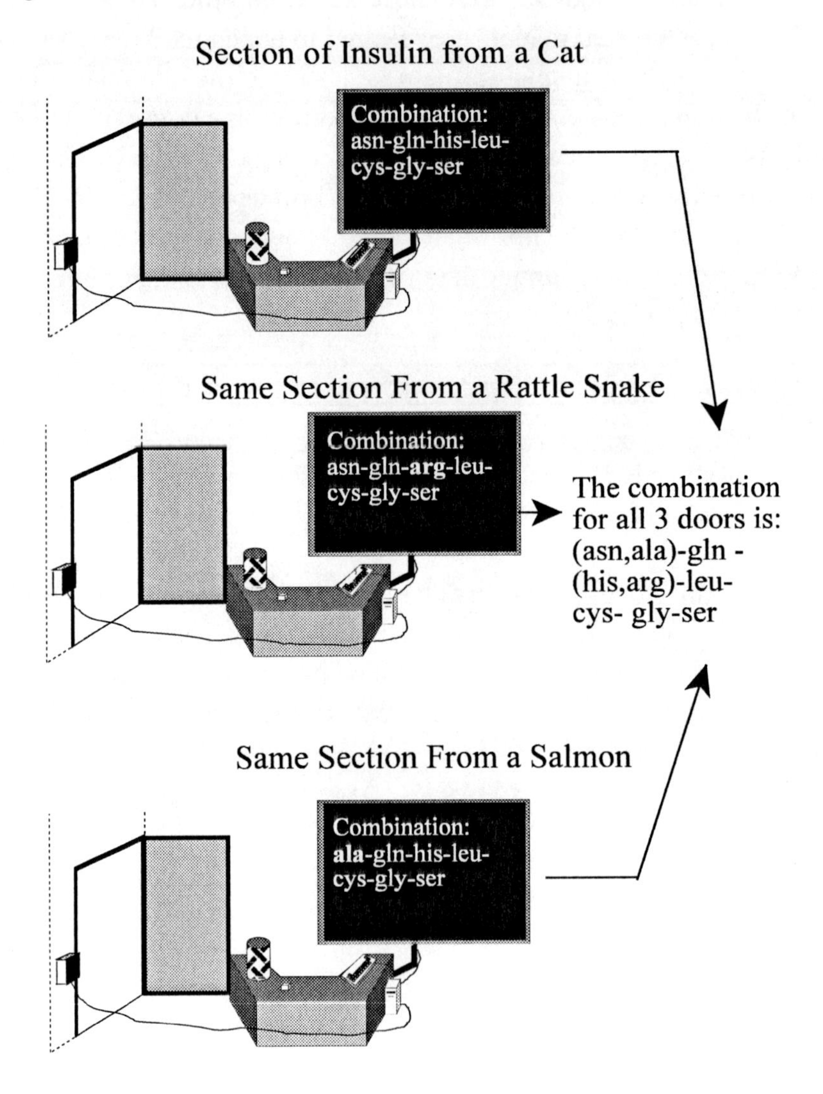

These 7 amino acid combinations corresponds to a section of the real insulin protein in cat, snake and fish. The differences are important in determining the allowed amino acids. The first amino acid is asparagine (asn) in rattlesnakes and cats, but it is alanine in salmon; thus, both ala and asn are acceptable at position number 1. At position 3, histidine (his) is found in both cats and fish, but arginine (arg) is found in snakes; thus, both his and arg are acceptable at position number 3. In figure 4.1, positions where more than one amino acid is allowed are represented by placing the allowed amino acids in parenthesis; therefore, position 1 is represented by (asn, ala) which means that either amino acid is fine at this position. From a comparison just like this, it is possible to determine which amino acids are allowed at every position in the A and B chains of insulin, and this will determine the amount of information contained in insulin today.

How Much Information Opens the Door

In figure 4.1, the composite combination, (asn, ala)-gln-(his, arg)-leu-cys-gly-ser, is assumed to be functional. In other words, this combination will open all three doors in figure 4.1. To compute the information in this composite combination, the odds of each amino acid arising by chance at each position must be known.

Table 4.1 - Odds of Each Amino Acid Arising by Chance

Number of Codons	Odds	Amino Acids
6	6 in 64	ser, arg, leu
4	4 in 64	val, pro, thr, ala, gly
3	3 in 64	ile
2	2 in 64	phe, tyr, his, gln, asn, lys, asp, glu, cys
1	1 in 64	met, trp

The odds of each amino acid arising by chance are listed in table 4.1. For example, the amino acid, serine (ser), will arise by chance 6 times in every 64 tries. The same is true for leucine(leu) and arginine(arg). The amino acid, methionine (met), will only arise by chance 1 time in 64 tries. Table 4.1 was created from table 3.2 by assuming that all mutations are random.

Table 4.1 will now be used to calculate the information in figure 4.1. In figure 4.1, the required combination is as follows:

pos1	pos2	pos3	pos4	pos5	pos6	pos7
asn	gln	his	leu	cys	gly	ser
ala		arg				

Position 1: asn (odds = 2 in 64) and ala (odds = 4 in 64) are both allowed. The odds of seeing either an asn or ala are found by simple addition. 2/64+4/64=6/64. In other words, the odds of an ala or asn arising by chance at position one are 6 times in 64 tries. Using the second equation presented in chapter 1, Information = 3.32xlog(64/6) = 3.4 bits. Position 1 contains 3.4 bits of information.

Position 2: gln has a 2 in 64 chance of arising by chance. This is equivalent to 1 chance in 32 tries. So the information is easy to find: $2^{(\text{information})}$ = 32/1. Since 2^5 =32, position 2 must contain 5 bits of information. Notice that logarithms can also be used. Information = 3.32 x log(32/1) = 5 bits. Both equations give the same result.

Position 3: The odds for histidine (his) are 2 in 64. The odds for arginine (arg) are 6 in 64. The sum determines the odds that one of these will be present. The sum is 8 in 64 which is equivalent to 1 in 8. The information is $2^{(\text{information})}$ = 8/1. Since 2^3=8, position three contributes 3 bits of information.

Position 4: Leucine has a 6 in 64 chance of arising by chance. These odds are the same as position 1. So position 4 also contributes 3.4 bits.

Position 5: cysteine (cys) has a 2 in 64 chance of arising by chance. These odds are the same as those calculated for position 2. So position 5 contributes 5 bits.

Position 6: glycine (gly) has a 4 in 64 chance of arising by chance. This is equivalent to 1 in 16. So the information at position 6 is $2^{(\text{information})} = 16/1$. Since $2^4=16$, position 6 contributes 4 bits of information.

Position 7: serine has a 6 in 64 chance of arising by chance. These odds are the same as those for position 4. So position 7 contributes 3.4 bits of information.

The total information required to open the door is the sum of the information found at each position or $3.4+5+3+3.4+5+4+3.4 = 27.2$ bits. The odds that this door can be opened by chance are given by 1 in $2^{27.2}$ or 1 in 154 million.

But the above calculation is wrong! The previous paragraph assigns a probability to a protein evolving based on its information content today. It completely ignores the ability of natural selection to guide evolution. While the information calculated is correct, 27.2 bits, this information has no relationship to the probability of any protein evolving. In figure 4.1, only the last door for each of the three species is shown. There may be many doors leading up to these doors, and the odds for success may be quite good. Information should never be related to a probability when natural selection is involved.

Why Does this Work? Common Ancestors

The evolution of all modern proteins can be traced back to a common ancestor. Figure 4.2 illustrates this concept. Around 500 million years ago suppose that there was a common ancestor for flounders and rabbits. Some of the descendants of this ancestor evolved into rabbits. Others evolved into flounders. Today the DNA of this common ancestor is not available, but the DNA of flounders and rabbits is certainly available. When the DNA of a flounder and a rabbit are compared, most of the information found in their DNA is the same. The insulin found in a flounder is very similar to that found in a rabbit, but there are differences because the two species have had 500 million years to accumulate changes independently.

Figure 4.2: Common Ancestors

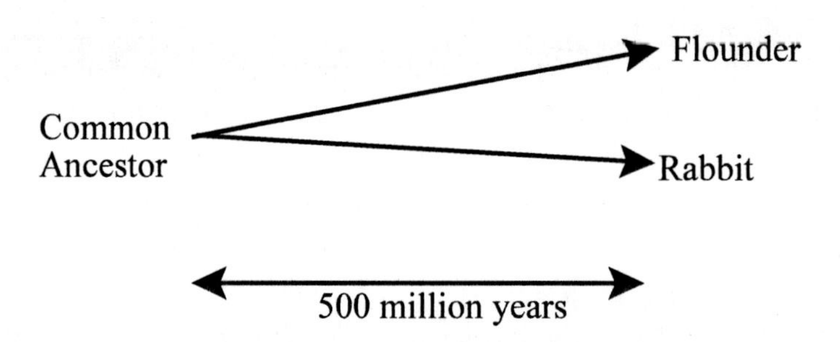

If a mutation modifies the insulin amino acid sequence, several fates exist for the modified protein.

- If the modified protein is better than the original, natural selection may encourage it to spread through the population. With time the new protein may be fixed in the population. This means that every member of the population possesses the modified protein.

- If the modified protein provides no selective advantage, it may still be fixed in the population. As long as the modified protein is as good as the original, but no better, the probability of fixation is equal to the rate of change.[3] Any protein that meets these criteria is termed neutral. So if a specific amino acid in insulin mutates every 100 million years, then a modified insulin with the changed amino acid is expected to be fixed in the population every 100 million years.

- If the mutation is slightly harmful, natural selection will most likely eliminate it from the population but not always.[3]

Assume for a minute that the amino acid sequence of insulin is not important and that almost any protein composed of 50 amino acids or more can signal cells to absorb sugar. In other words, the insulin hormone contains almost no useful information. If this assumption is true, then one would expect the insulin amino acid sequence in fish and in mammals to be completely different. The sequences have had 500 million years to change independently.

Analysis of insulin in fish and in mammals has revealed that this is not the case. Many of the amino acids are the same or have similar chemical properties. These amino acids are said to be conserved.

Conserved amino acids are a measure of information. To accurately measure this information, a comparison of many diverse species is required. The more diverse the species the better. The technique works best for proteins that are found in all of the kingdoms of life.

Total Information in Insulin A and B Chains

The techniques used in the previous section can easily be extended to calculate the total information in insulin. Figure 4.3 was generated using a software package designed to align the amino acid sequences of similar proteins, Clustal X. Instead of using the three letter abbreviation for each amino acid, the single letter abbreviation is used to conserve space. The letters across the first row are the amino acid sequence of insulin in chickens, the second row snakes, and the last row flounders. The columns are aligned by the Clustal X program so that similar amino acids appear in the same column. The dashes represent gaps inserted by the Clustal program to align the sequences.

Figure 4.3: The Amino Acid Sequence of Insulin

CLUSTAL X (1.8) MULTIPLE SEQUENCE ALIGNMENT

File: C:clustalinsulin1.ps **Date: Wed Sep 15 08:54:24 2004**
Page 1 of 1

```
           *:***.***:**:*:**:***:* *: ***:***   *.:::*: ***
 Chicken -AANQHLCGSHLVEALYLVCGERGFFYTPKAGIVEQCCHNTCSLYQLENYCN     51
   Snake -APNQRLCGSHLVEALFLICGERGFYYSPRSGIVEQCCENTCSLYQLENYCN     51
    Goat -FVNQHLCGSHLVEALYLVCGERGFFYTPKAGIVEQCCAGVCSLYQLENYCN     51
   Sheep -FVNQHLCGSHLVEALYLVCGERGFFYTPKAGIVEQCCAGVCSLYQLENYCN     51
     Cat -FVNQHLCGSHLVEALYLVCGERGFFYTPKAGIVEQCCASVCSLYQLEHYCN     51
Elephant -FVNQHLCGSHLVEALYLVCGERGFFYTPKTGIVEQCCTGVCSLYQLENYCN     51
     Dog -FVNQHLCGSHLVEALYLVCGERGFFYTPKAGIVEQCCTSICSLYQLENYCN     51
     Pig -FVNQHLCGSHLVEALYLVCGERGFFYTPKAGIVEQCCTSICSLYQLENYCN     51
   Whale -FVNQHLCGSHLVEALYLVCGERGFFYTPKAGIVEQCCTSICSLYQLENYCN     51
     Rat -FVKQHLCGSHLVEALYLVCGERGFFYTPKSGIVDQCCTSICSLYQLENYCN     51
   Mouse -FVKQHLCGPHLVEALYLVCGERGFFYTPKSGIVDQCCTSICSLYQLENYCN     51
 Hamster -FVNQHLCGSHLVEALYLVCGERGFFYTPKSGIVDQCCTSICSLYQLENYCN     51
   Rabit -FVNQHLCGSHLVEALYLVCGERGFFYTPKSGIVEQCCTSICSLYQLENYCN     51
   Human -FVNQHLCGSHLVEALYLVCGERGFFYTPKTGIVEQCCTSICSLYQLENYCN     51
   Horse -FVNQHLCGSHLVEALYLVCGERGFFYTPKAGIVEQCCTGICSLYQLENYCN     51
    Frog -LVNQHLCGSHLVEALYLVCGDRGFFYYPKVGIVEQCCHSTCSLFQLESYCN     51
Toadfish MAPPQHLCGSHLVDALYLVCGDRGFFYNPK-GIVEQCCHRPCDIFDLQSYCN     51
Flounder VVPPQHLCGAHLVDALYLVCGERGFFYTPK-GIVEQCCHKPCNIFDLQNYCN     51
   ruler 1.......10........20........30........40........50..
```

Single letter amino acid abbreviations used in figure 4.3:

A = Ala	C = Cys	D = Asp	E = Glu
F = Phe	G = Gly	H = His	I = Ile
K = Lys	L = Leu	M = Met	N = Asn
P = Pro	Q = Gln	R = Arg	S = Ser
T = Thr	V = Val	W = Trp	Y = Tyr

Insulin is composed of two chains, A and B. Referring to the ruler at the bottom of figure 4.3, the B chain is comprised of amino acids 2-31, and the A chain is comprised of amino acids 32-52. The columns are aligned in such a way to match up the amino acids when they are the same. For example, the last amino acid on each row is always N. This means that N (the amino acid Asparagine or Asn) is conserved. Notice that even in columns where the amino acids differ, the variability is still quite low. Insulin is a highly conserved protein.

Tables 4.2 and 4.3 calculate the information in insulin as it exists today. To help with understanding, a few sample calculations are described first.

Example calculation for table 4.2: refer to figure 4.3, and find the 10th column. Notice that it contains A (alanine) in flounders, S (serine) for most species, and P (proline) in mice. This means that all three of these amino acids are acceptable at this position. The information content is calculated as follows (refer to table 4.1): information = 3.32 x log[64 possible outcomes/ (4+6+4) observed outcomes] = 2.2 bits. Thus, row 10 in table 4.2 is assigned 2.2 bits.

Example calculation for table 4.3: refer to figure 4.3, column 42. This column is always C (cysteine), thus, the information is as follows: information = 3.32 x log [64/2] = 5 bits. The total information for each chain is the sum of the information at each position. The sum of columns 3 and 6 in table 4.3 is 81 bits.

Table 4.2: Information in Insulin (B chain only)

pos	allowed amino acids	bits	pos	allowed amino acids	bits
2	phe, ala, leu, val	2.0	17	phe, tyr	4
3	val, ala, pro	2.4	18	leu	3.4
4	pro, lys, asn	3	19	val, ile	3.2
5	gln	5	20	cys	5
6	his, arg	3	21	gly	4
7	leu	3.4	22	asp, glu	4
8	cys	5	23	arg	3.4
9	gly	4	24	gly	4
10	ala, pro, ser	2.2	25	phe	5
11	his	5	26	phe, tyr	4
12	leu	3.4	27	tyr	5
13	val	4	28	thr, ser, asn	2.4
14	glu, asp	4	29	pro	4
15	ala	4	30	lys, arg	2.7
16	leu	3.4	31	ala, thr, ser, -	0

Total = 108 bits

Table 4.3: Information in Insulin (A chain only)

pos	allowed amino acids	bits	pos	allowed amino acids	bits
32	gly	4	43	ser, asn, asp	2.7
33	ile	4.4	44	leu, ile	2.8
34	val	4	45	phe, tyr	4
35	glu, asp	4	46	gln, asp	4
36	gln	5	47	leu	3.4
37	cys	5	48	glu, gln	4
38	cys	5	49	asn, ser,his	2.7
39	glu, his, thr, ala	2.4	50	tyr	5
40	asn, lys, arg, ser, gly	1.67	51	cys	5
41	pro,thr,ile,val	2.1	52	asn	5
42	cys	5			

Total = 81 bits

The total information today in insulin is the sum of the information found in both chains or 81 +108 = 189 bits. It is important to keep in mind that this number has absolutely nothing to do with the probability of insulin evolving.

Molecular Knowledge of Insulin

Finding the information contained in insulin is straight forward. The math is tedious, but the procedure is at least defined, and today insulin contains 189 bits of information. So how much of this information does insulin require to provide a selective advantage? This question is much more difficult to answer.

This is where human insight is necessary. Some amino acids side chains have very similar chemical properties. Others are similar in size. Thus, some amino acid substitutions should be allowed even if they are not found. These are summarized below with the chemical trait given in parentheses:

Group1: leucine , isoleucine, valine, alanine, and methionine (do not like water, so they tend to cluster on the inside of the protein).
Group 2: tyrosine, phenylalanine, and tryptophan (very large amino acids that can influence protein folding).
Group 3: aspartate and glutamate (acidic side chains, like water).
group 4: histidine, arginine, and lysine (basic side chains, like water).
Group 5: glutamine and asparagine (charged and like water).
Group 6: serine and threonine (like water, tend to be found on the outside of protein).
Group 7: glycine (very small).
Group 8: proline (introduces a bend into the chain).
Group 9: cysteine (cross links peptide chains).

Based on these properties, this chapter will propose the following procedure to calculate knowledge: if a column in a multiple alignment sequence like figure 4.3 only contains a single amino acid, or if the variation is limited to any one of the above 9 groups, then the column should be included in the calculation for molecular knowledge. If the column contains amino acids from different groups then it should be excluded.

Furthermore, for the columns included in the molecular knowledge calculation, all amino acids in the same group must be included whether they are present in the alignment or not. For example, at position 19 in table 4.2, only isoleucine and valine are found. But because alanine, methionine, and leucine belong to group 1, it is assumed that these amino acids can be substituted at position 19 without destroying the function of insulin. With this procedure, table 4.2 becomes table 4.4. The parenthesis in table 4.4 represent amino acids that are not present in the multiple sequence alignment (figure 4.3). The positions that are assigned 0 bits have amino acids from more than one of the 9 predefined groups.

Table 4.4: Molecular Knowledge in B chain of Insulin

pos	allowed amino acids	bits	pos	allowed amino acids	bits
2	phe, ala, leu, val	0	17	phe, tyr ,(trp)	3.7
3	val, ala, pro	0	18	leu, (ile),(val), (ala), (met)	1.8
4	pro, lys, asn	0	19	val, ile, (ala), (leu), (met)	1.8
5	gln, (asn)	4	20	cys	5
6	his, arg, (lys)	2.7	21	gly	4
7	leu, (ile), (leu), (val), (met)	1.8	22	asp, glu	4
8	cys	5	23	arg, (lys), (his)	2.7
9	gly	4	24	gly	4
10	ala, pro, ser	0	25	phe, (tyr), (trp)	3.7
11	his, (lys), (arg)	2.4	26	phe, tyr, (trp)	3.7
12	leu, (ile), (val), (ala), (met)	1.8	27	tyr, (phe), (trp)	3.7
13	val, (ile), (leu), (ala), (met)	1.8	28	thr, ser, asn	0
14	glu, asp	4	29	pro	4
15	ala, (leu), (ile),(val), (met)	1.8	30	lys, arg, (his)	2.4
16	leu, (ala), (val),(Ile), (met)	1.8	31	ala, thr, ser, -	0

Total = 76 bits

Example calculation: at position 3 val, ala and pro are found. Because these amino acids are in different groups, the knowledge is defined as zero bits. At position 16 only leu is found, but ala, val, ile, and met probably will not be that damaging to protein function because they are in the same group. The total number of codons that encode these 5 amino acids is 18. Thus, knowledge = 3.32 x log[64/18] = 1.8 bits.

Comparing table 4.4 to table 4.2, it is clear that knowledge is much less than information (76 bits vs. 108 bits). The ratio of knowledge to information for the insulin B chain is thus 76/108 = 70%. The same procedure is repeated for the A chain as shown in table 4.5.

Table 4.5: Molecular Knowledge in Insulin A Chain

pos	allowed amino acids	bits	pos	allowed amino acids	bits
32	gly	4	43	ser, asn, asp	0
33	ile, (val), (leu), (ala), (met)	1.8	44	leu, ile, (val), (ala), (met)	1.8
34	val, (leu), (ile), (ala), (met)	1.8	45	phe, tyr, (trp)	3.7
35	glu, asp	4	46	gln, asp	0
36	gln, (asn)	4	47	leu, (val), (ile), (ala), (met)	1.8
37	cys	5	48	glu, gln	0
38	cys	5	49	asn, ser,his	0
39	glu, his, thr, ala	0	50	tyr, (phe), (trp)	3.7
40	asn, lys, arg, ser, gly	0	51	cys	5
41	pro,thr,ile,val	0	52	asn, (gln)	4
42	cys	5			

Total = 51 bits

Cartoon and Space Fill Models of Insulin

The following pictures of the A and B chains show the location of the amino acids that contain information. The following color code applies: black> 3.4 bits, dark gray> 2.5 bits, gray>2 bits, light gray >1 bit, white <1 bit.

Figure 4.4: Space fill and Cartoon model of B Chain

Figure 4.5: Space fill and Cartoon model of A Chain

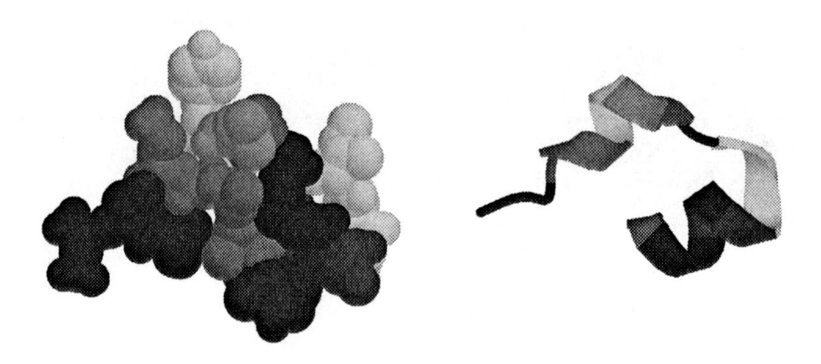

Figure 4.6: Space Fill and Cartoon model of Entire Molecule

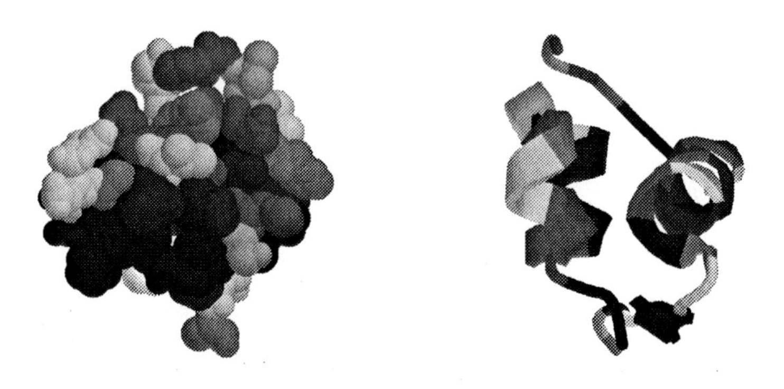

The upper right front cover picture also represents insulin. This picture is in color. Red sections contribute > 4 bits per amino acid, orange > 3.4 bits per amino acid, yellow > 2.5 bits per amino acid, and green > 2 bits per amino acid.

The Probability of Insulin Evolving

The calculation for the amount of information in the insulin A and B chain is based strictly on mathematics. No interpretation is required. The same cannot be said for molecular knowledge. Molecular knowledge requires finding the minimum information that preserves some functionality.

It is a mistake to calculate the odds for the evolution of insulin based on the information in insulin. It is extremely important to use the minimum information that results in a protein with some selective advantage. In other words, molecular knowledge must be used to calculate the probability associated with a protein evolving, and molecular knowledge can only be found by using math along with human insight.

In this chapter, the information in insulin is found to be 189 bits, and the molecular knowledge is calculated at 127 bits. Figure 4.7 illustrates the steps that chance must overcome.

Figure 4.7: Molecular Knowledge in Insulin

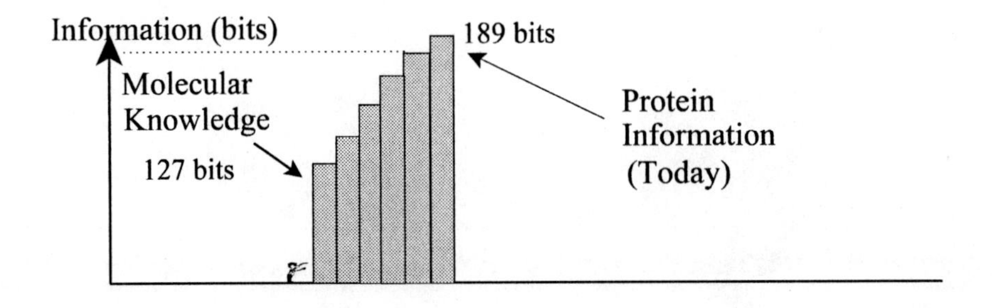

Insulin May Not Imply Design

In most cases, the odds that a protein can evolve are simply 1 in $2^{(\text{molecular knowledge})}$, or in this case, 1 in 2^{127}, but this technique may not apply to insulin. Insulin binds to a protein called the insulin receptor. This receptor senses insulin, and through a few more steps signals cells to absorb sugar from the bloodstream.

This receptor is very specific to the insulin hormone. The original receptor may have been much less specific. So while today insulin requires 127 bits of knowledge, a precursor that might have existed 500 million years ago may have required much less. One could certainly envision a very different insulin receptor. Perhaps this receptor signaled the cell to absorb sugar when it detected any protein greater than 20 amino acids. In this case, the first insulin molecules would have required almost no molecular knowledge, and figure 4.7 might look like figure 4.8.

Figure 4.8: Molecular knowledge in Insulin

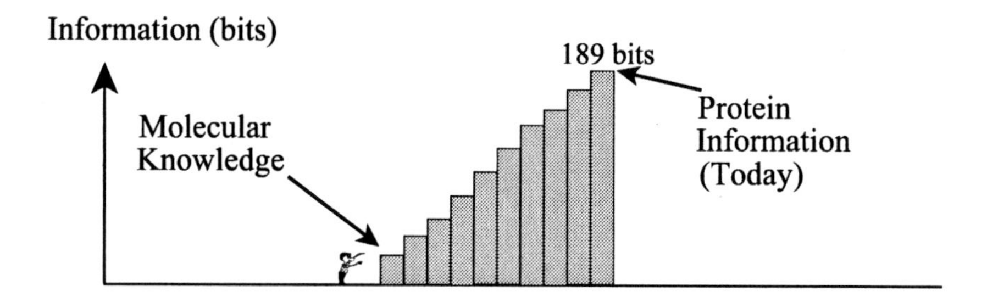

If figure 4.8 is accurate then there is a very clear path for Darwinian evolution to work just like Darwin theorized. All of the steps are small; thus, the scientist can easily climb to the top.

Because the structure and specificity of the first insulin receptor is unknown, there is no way to choose between figure 4.7 and 4.8; therefore, the molecular knowledge in insulin cannot be used to reliably infer design. There is no way to choose between figure 4.7 and figure 4.8. Insulin was chosen because it is a very small protein. This makes it easy to manually calculate the information and hence the knowledge. Insulin was not chosen because it implies design. It is merely a convenient learning tool.

Insulin is unique in that its required molecular knowledge depends on its receptor. This allows insulin to co-evolve with its receptor. Most proteins do not have this option.

While figure 4.8 cannot be ruled out, if this figure is an accurate representation, then it can only represent the evolution of insulin before the existence of the common ancestor to fish, amphibians, reptiles, birds and mammals. This must be true because of the similarity of the insulin protein in these diverse species today. This similarity is very common, and it is observed in most proteins. Once proteins exist, they tend not to evolve (see chapter 16).

How Accurate is the Technique?

Molecular knowledge has no mathematical definition. Relying on human insight to find it opens the door to different interpretations. Nevertheless, the technique presented in this chapter to calculate molecular knowledge will always underestimate the true value. For example, at position 24 in table 4.5 only valine is found. With this technique, leucine, isoleucine, methionine and alanine are also allowed because they belong to the same amino acid group. If this valine is replaced with leucine, the resulting insulin will lose more than 90% of its functionality.[2] Replacing it with an amino acid from a different group should further destroy functionality (leaving the insulin molecule with almost no functionality). Positions 25 and 26 of the B chain are always tyrosine or phenylalanine. Replacing position 26 with a serine almost completely destroys insulin functionality.[1] Position 25 is not as critical because a serine here only destroys 85% of insulin's functionality.[1] In the case of insulin, functionality is a measure of how effective the modified insulin molecule signals cells to absorb sugar compared to unmodified insulin. Molecular knowledge implies that the protein has some minimal functionality. Changing a single conserved amino acid can radically impair insulin's function. The same is true for most proteins.

Because the technique used to calculate molecular knowledge in this chapter allows many amino acid substitutions, it classifies billions and billions of proteins that are expected to have no function as functional. While no technique to calculate molecular knowledge will be perfect, the one introduced in this chapter will always underestimate the true value of molecular knowledge. In many cases, this technique will assign 0 bits to most of the amino acids in a protein (see chapter 14 calculations for the protein G3PD); as a result, this technique often grossly underestimates the actual knowledge in proteins.

The technique introduced here is not meant to be used as a tool to predict which amino acids can be changed without destroying protein function. In this model, many amino acids that are not allowed are classified as functional. For instance, at positions, 39, 40, 41, 43, 46, 48, and 49 (table 4.5), every single amino acid is allowed, and a quick inspection of table 4.5 reveals that at these positions very few amino acids are found in a real insulin protein. Thus, it is quite likely that most of the amino acid sequences that this model classifies as functional are in fact not.

If molecular knowledge is plotted against amino acid number, then something similar to figure 4.9 might result.

Figure 4.9: Actual knowledge vs. Assigned Knowledge

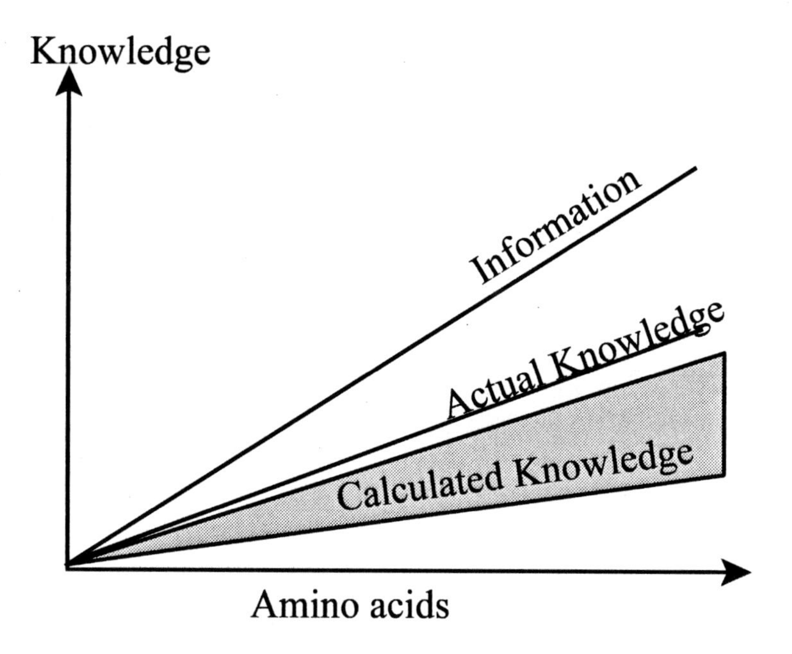

If the procedure introduced in this chapter classifies a few functional proteins as nonfunctional, then this will tend to raise the calculated knowledge line in figure 4.9. Conversely, if this technique classifies some proteins that are nonfunctional as functional, then the line for calculated knowledge is depressed. If and when the two errors exactly balance, then the calculated knowledge will equal the actual knowledge. This is unlikely because the technique used to calculate knowledge classifies many more nonfunctional proteins as functional than vice versa. This always depresses the calculated knowledge. Because the magnitude of this depression depends on the protein being analyzed, the calculated knowledge is shown as a range of values. Nevertheless, the calculated knowledge is always less than the actual knowledge, so when the calculated knowledge is related to a probability, the odds for success will always be overly optimistic.

Figure 4.10 on the next page should be compared to figure 4.1 at the beginning of this chapter. In figure 4.1, the technique used to calculate the information of insulin (as it exists today) is shown. This information cannot be related to a probability because insulin has already been optimized by natural selection. In figure 4.1, the only door is the last door. All of the doors leading up to this door are hidden. The technique introduced in this chapter attempts to reconstruct the combination of the earlier doors. In particular, the first door is important because it is this door that determines whether or not naturalistic laws can explain the evolution of insulin. The screen in figure 4.10 shows the combinations that will open the first door. For example, the first position is represented by an asterisk because all 20 amino acids are allowed at this position. In the second position, only gln is found today, but since asn belongs to the same group, asn is shown in parenthesis. Any combination with either gln or asn at the second position will open the door. The accuracy of this technique depends on how well it predicts the combination of the first door. Insulin may not imply design for reasons already discussed. Nevertheless, when other proteins are analyzed with this method, the design inference is very strong.

Figure 4.10: Molecular Knowledge of Insulin

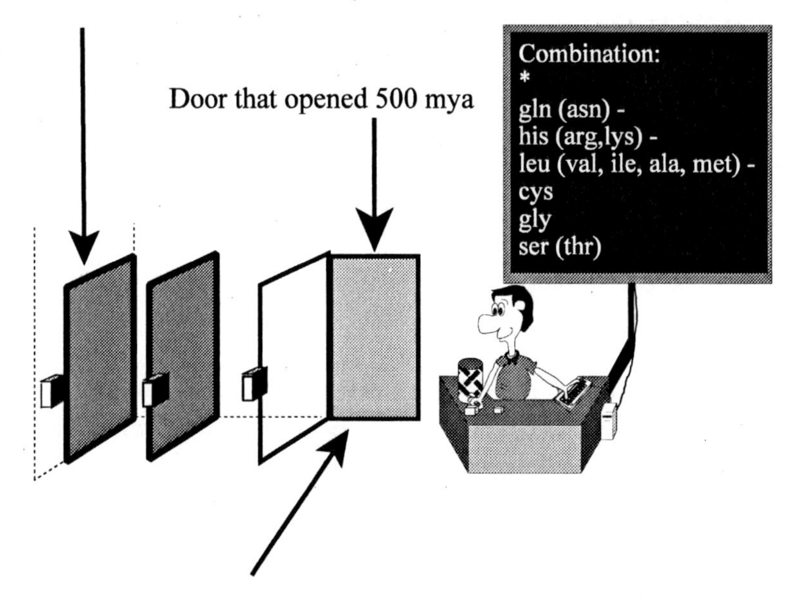

References:

1) Shoelson et al.,"Identification of a Mutant Human Insulin Predicted to Contain a Serine for Phenylalanine Substitution," Procedding of the National Association for Science, Dec 1983, vol 80, 7390 -7390.

2) Sakura H., et al, "Structurally Abnormal Insulin in a diabetic Patient Characteristic of the Mutant Insulin A3 (Val-> leu) Isolates from the Pancreas," J. Clin. Invest. 78:1666-1672, 1986.

3) The Neutral Theory of Evolution, Kimura, 1983.

4) Thompson, Gibson, Plewniak, Jeanmougin and Higgins, The ClustalX windows interface: flexible strategies for multiple sequence alignment aided by quality analysis tools. Nucleic Acids Research, 24:4876-4882, 1997.

Links from this web site www.evolution-by-design.com offer interactive pictures of insulin whose amino acids can be colored by information and knowledge. Viewing these pictures may require a high speed internet connection and a reasonably fast computer.

Supplemental Material:

The amino acid sequences of insulin and almost all other proteins found in life are available at http://us.expasy.org/sprot/. A search for insulin on this web site returns the sequence for insulin in many different species. Browse down the search results to locate INS_HUMAN. This link contains the sequence for human insulin (at the very bottom of the page). Notice that the amino acid sequence contains 110 amino acids. After the initial insulin peptide is created, 59 amino acids are removed to create the final protein, which contains the A and B chains. The amino acid sequence of the A and B chain is also available above the 110 amino acid sequence. Several variants of the normal sequence are also listed in this section. These variations are usually associated with diabetes.

Figure 4.3 was constructed by downloading the insulin amino acid sequence for 18 species in FASTA format, and then running the alignment program, ClustalX, which is available at this URL: http://www-igbmc.u-strasbg.fr/BioInfo/ClustalX/Top.html

The sample size was limited to 18 species because the online databases are not without error. The criteria to include a position in molecular knowledge is very sensitive to these errors. For example, suppose that position 10 in a specific protein is always a glycine, but due to an error in the database, one of the entries reports that a valine is found at this position instead of glycine. This error means that position 10 must be excluded from the calculation in molecular knowledge. So to make this technique less sensitive to database errors, the sample size is arbitrarily limited to 18 entries.

Part 2: Chemical Evolution

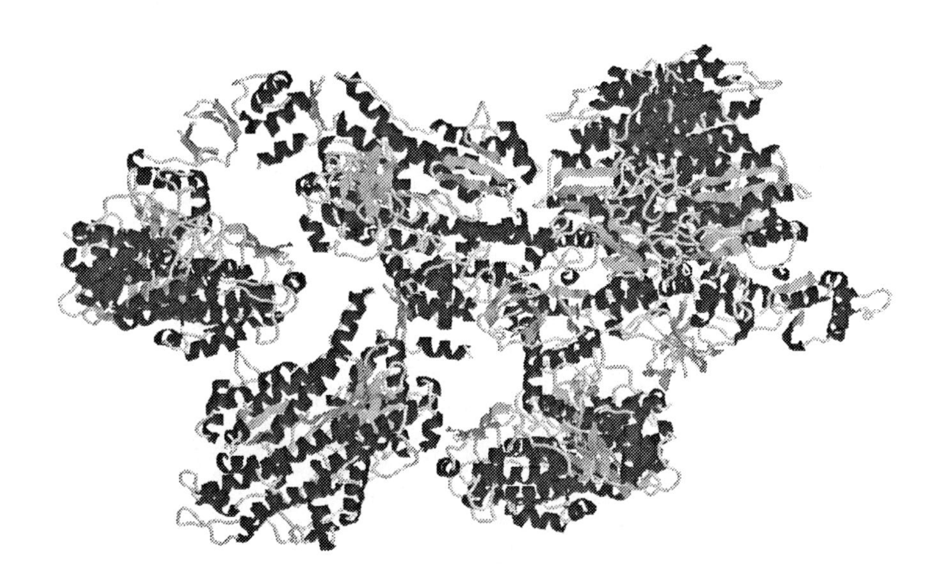

The figure is a cartoon representation of the protein Myosin. Myosin and actin are the two proteins responsible for muscle contraction.

Chapter 5: Information & Knowledge before the Genetic Code

To calculate the information and knowledge for insulin in the last chapter, the genetic code was used to assign a probability to each amino acid arising by chance. For this calculation to be meaningful, both the code and a method to turn the knowledge in DNA into proteins must already exist. So the calculations in chapter 4 assume that life already exists. What about before life exists? How would one calculate the information or knowledge in the very first protein? This is not an easy problem.

Several authors have used thermodynamics, but thermodynamics only applies when the system reaches equilibrium. The relevance of thermodynamic calculations is questionable as amino acids do not polymerize into peptides chains unless external conditions force them away from equilibrium.

This chapter will use information theory to solve the problem. Unlike thermodynamics, information theory can easily deal with non-equilibrium systems.

Information theory cannot normally be used to predict how chemicals will react because some chemicals react with each other readily, and others only react very slowly. Others do not react with each other at all. Thus, the likelihood of two chemicals joining together depends on both the quantity of the chemicals present and their chemical properties. Information theory can easily deal with the effects of quantity, but it has no way to deal with chemical properties.

This chapter will require several assumptions. Without these assumptions information theory cannot be applied to chemical reactions. Fortunately, these assumptions will improve the probability for creating a protein in the primordial soup.

Assumptions:

- The probability of a peptide bond forming between two amino acids only depends on how many of each amino acid is present in the system.
- The primordial soup only contains amino acids.
- Amino acids do not form non-proteinous bonds with each other. So for example, the carboxylic acid functional groups in aspartate and glutamate do not react with the n-terminus of other amino acids.

The first assumption allows all amino acids to be treated equally. While this assumption ignores the chemical properties of each amino acid, the assumption is not an unreasonable approximation because all amino acids must join together by forming a peptide bond. This assumption improves the odds because it ignores the finding that glycine and alanine are not only the most common amino acids but they are also the most likely to form the alpha peptide bonds required by proteins (Fox, 1972, page144 and 154). The second assumption also greatly improves the odds of creating a functional protein. By excluding chemicals that react quickly with amino acids, this assumption eliminates chemical reactions that can prematurely terminate a growing peptide chain. It also ensures that the amino acids will be available to interact with each other. The third assumption is not true, but it greatly simplifies the math, and at the same time, it improves the odds of creating a protein in the soup.

With these assumptions, information theory may be applied to the primordial soup. The first step is to estimate the number of each amino acid in the primordial soup. There are two methods. For 50 years, scientists have been trying to find better ways to synthesize amino acids under plausible prebiotic conditions. Many of the 20 amino acids used by life have been synthesized. Because these experiments are riddled with speculation about conditions on the primitive earth and investigator interference, the second method is preferable. This method relies on the amino acids found in meteorites.

Meteorites

Some meteorites contain organic carbon, and several of these have been analyzed for amino acids. This analysis has shown that the amino acids, glycine and alanine, are quite common in some meteorites. Most of the other amino acids used by life are rare, but some are present. In addition, many amino acids not used by life are present. More than 50 non-biological amino acids are found in the Murchison meteorite.

Meteorites are easily contaminated by biological amino acids. So samples are always taken from the meteorite interior. Unfortunately, contamination is still a major problem. Nevertheless, several generalizations are possible.

- The biologically relevant amino acids in meteorites are always predominantly glycine and alanine. Sometimes aspartate and glutamate run a close second, but in many cases, this appears to be the result of contamination. Serine and valine are sometimes present. The other amino acids used by life are absent.

- Non-biological amino acids are common in variety and in number. The most common non-biological amino acids are the many isomers of aminobutyric acid. The second most common non-biological amino acids are two forms of alanine that life does not use.

One comparison of four different meteorites that contain amino acids revealed that only 25% of the amino acids are biologically relevant.[4] If the primordial soup has a similar composition, then only 25% of the amino acids in the soup are biologically relevant, and even if a way is found to make the amino acids join together, the odds of a protein emerging are very small.

The average protein in the Swiss Prot database contains 362 amino acids, and most contain more than 150 amino acids. If the composition of amino acids in the soup is similar to that of meteorites, what is the probability of creating a peptide composed of 150 amino acids if all of the amino acids must be biological?

The knowledge required to build this peptide is simply the knowledge required to exclude all amino acids not used by life. Today, random amino acid sequences do not contain such knowledge because the machinery used by life to build proteins ensures that one of the 20 amino acids used by life will always be placed at each position in the growing chain. This is not true in the primordial soup. The term molecular knowledge in this book is reserved for useful information that conveys a selective advantage. A random sequence of biological amino acids that evolves in the soup will not possess molecular knowledge because the sequence will most likely have no function. Thus, the term information is preferred in this case. To avoid any possible confusion with terminology, this type of information will always be referred to as primordial information. Primordial information is the information needed to exclude non-biological chemicals found in the primordial soup from a growing peptide, RNA or DNA molecule. Since primordial information is a form of knowledge, it can be safely related to a probability. Furthermore, this calculation does not rely on human insight. Before self replication, natural selection cannot exist, so all events are guided by chance and chance alone.

Each addition to the growing chain has a 25% chance of being an amino acid used by life. So each amino acid added to the chain has a 1 in 4 chance of being correct. Thus, there are 4 possible outcomes and only one is desirable. Using equation 1 in chapter 1, $2^{\text{information}} = 4/1$, and because $2^2 = 4$, the information content for each amino acid added is 2 bits. So a random chain of 150 amino acids that emerges from the soup will contain 300 bits of information. The odds of this arising by chance are 1 time in 2^{300} tries or a 1 in 2 x 10^{90} chance. What do odds like 1 in 2 x 10^{90} really mean?

The number 10^{90} is so large that naturalistic explanations will always fail to explain any event whose odds are this poor. To understand why, assume that every single star in the universe has one planet composed entirely of amino acids. Further assume that every one of these amino acids exists as a 150 amino acid peptide chain. The highest estimate for the number of stars in the universe currently available is 7×10^{22}. If the planets orbiting these 7×10^{22} stars are about the same size as the earth, then on average each has a mass of 6×10^{24} Kg. A planet with this mass composed entirely of the amino acid glycine will be made from 5×10^{49} glycine molecules. If all the planets have the same number of amino acids, then there will be 3.5×10^{72} amino acids in the universe. Since every amino acid exists in a chain of 150, there will be 2.3×10^{70} peptide chains. The odds that 1 of these chains will contain only biologically relevant amino acids is only 1 in 8.6×10^{19}. So further assume, that all of these peptide chains break down each year only to reform, and that this process has been going on every year for 15 billion years. The odds improve to 1 in 6 billion. So while the odds are not zero, they might as well be. Nature simply cannot accumulate enough tries to overcome the poor odds.

One can certainly speculate that the first proteins used amino acids that are no longer used today or that these proteins were very short. Both assumptions improve the likelihood for evolution. Nevertheless, all readers need to realize that when a scientist in the lab mixes together pure amino acids that are only used by life, the scientist is adding so much information to the system that the experiment can no longer be considered representative of the conditions on the early earth. The starting point for such experiments is not plausible.

If the soup existed, then the first proteins evolved in a soup that contained many amino acids not used by life. The soup also contained a host of other chemicals like aldehydes that react readily with amino acids. These undesirable side reactions make the evolution of information in the primordial soup very difficult to explain. When a scientific experiment models evolution by excluding these other chemicals, the experiment no longer models the origin of life. Such experiments only model evolution in a test tube.

The Evolution of Primordial Knowledge

The odds of a random amino acid chain evolving in the soup are quite poor, but what about a protein? A protein is not a sequence of random amino acids. The order and type of amino acids in a protein determine how it folds, how it behaves, and its biological function. The sequences are not random. They contain knowledge. The odds of a functional protein evolving are certainly expected to be much less than that of a random sequence.

How Many Solutions?

One of the more important experiments concerning the origin of life was performed by Keefe and Szostak.[1] The authors of this paper in Nature searched six trillion random peptides each composed of 80 amino acids. They were looking for a sequence that could bind the chemical, ATP. They found four sequences in this large pool with ATP binding activity.

This allows for a direct computation of the molecular knowledge required for ATP binding. Using equation 2 in chapter 1, molecular knowledge = 3.32 x log (6 trillion/4) = 40 bits. Notice, that this is not 40 bits of information because the proteins that were selected only possessed minimal functionality. These proteins were subjected to several rounds of selection greatly improving their affinity for ATP.

This experiment provides a direct measurement of molecular knowledge. It also shows that there are very few solutions.

Binding a chemical like ATP is one of the functions that many enzymes possess. So while Keefe and Szostak did not actually find a useful enzyme, they did find a function than many enzymes require. The 40 bits calculated above are for evolution in a test tube. How many bits are required for evolution in the primordial soup?

The minimum possible primordial information in a random sequence of 80 amino acids is 160 bits (2 bits per amino acid). The odds of such a peptide evolving are one in 1.5×10^{48}. Given that the odds that a random sequence of 80 amino acids will bind ATP are only 4 in 6 trillion, the odds of finding a primitive peptide on the earth that can bind ATP are simply the product of the two numbers or one chance in 2.2×10^{60}. Alternatively, the 160 bits needed to construct an 80 amino acid peptide in the soup may be added to the 40 bits calculated above. The total knowledge is thus 200 bits, and the odds of this happening are 1 in 2^{200} or 1 in 2.2×10^{60}. In this calculation, the total knowledge required is simply the sum of the molecular knowledge and the primordial information. After life exists, primordial information always equals zero, and molecular knowledge always equals total knowledge.

Binding ATP is a simple function. Clay, a simple mineral, binds ATP. Furthermore, the function by itself does not confer a selective advantage. Thus, ATP binding is below the threshold of molecular knowledge. This function must be combined with another function before natural selection will preserve it. To create a functional enzyme that can be preserved by natural selection quite a bit more knowledge is required. Since it takes 200 bits to bind ATP, assume that it also takes 200 bits to bind another molecule. Thus, 400 bits is a more reasonable approximation for a functional enzyme, and the odds for such evolution are given by 1 time in 2^{400} tries or a 1 in 2.5×10^{120} chance.

The origin of the first enzyme just cannot be explained in this way. The odds are too poor.

Molecular Knowledge Before Life

This section will investigate how the composition of the soup influences knowledge. If the soup contains mostly glycine and alanine along with a host of other amino acids not used by life, then the probability of a useful protein emerging from it must be very low. Chapter 4 calculated the molecular knowledge for the protein insulin. This chapter will repeat this procedure assuming that insulin emerged in the soup before life. By this calculation this chapter does not mean to suggest that insulin originated in the soup. The calculation is for comparison only. Remember insulin was only chosen because it does not contain many amino acids, and this makes the calculations easier.

The Composition of the Soup

If meteorites are used to reconstruct the composition of the soup, then 14 of the 20 amino acids used by life will be absent. Only glycine, alanine, valine, serine, aspartate and glutamate would be available in the soup. The proteins used by life today require more than 6 amino acids. While this prediction of the soup's composition is probably the most accurate, it is an undesirable composition. So this chapter will assume a much more favorable composition.

Life uses 20 amino acids. Seventeen of these have been synthesized in the lab under conditions that might be similar to the conditions found on earth 4 billion years ago. Some amino acids are quite easy to synthesize, and others are very difficult. The amino acids that are easy to synthesize invariable are the primary product of these experiments. The other amino acids occur in various concentrations depending on the conditions chosen to carry out the experiment. Three amino acids, histidine, arginine, and lysine, have not been synthesized under plausible conditions.[2]

Because no single experiment has generated all of the amino acids, if the soup's composition is taken from the results of a single prebiotic experiment, then the composition will be unfavorable for protein evolution. Most proteins need 18 or 19 different amino acids to function. To construct a favorable composition for protein evolution, it is either necessary to combine many different prebiotic experiments or to just assume that the absent amino acids are present. This section will take the latter approach.

On page 87 of his book, Miller lists the results from one of the most successful prebiotic experiments.[3] The yields of ten amino acids are listed in this table.

As a reasonable starting point, assume the abundance of the amino acids in the primordial soup tracks Miller's table. Ten amino acids are not found in Miller's table. Seven of these have been synthesized under plausible prebiotic conditions. Assume that these seven are as abundant as threonine. Threonine is the least common amino acid listed in Miller's table. Three amino acids have not been synthesized in the lab. Assume that these are found in the soup at 1/10 the concentration of threonine. Finally, assume that the 20 amino acids that life uses comprise 1/4 of all amino acids present in the soup. Thus, the soup ratio of biological to non-biological amino acids is similar to the ratio found in meteorites.

These assumptions improve the odds that a protein will emerge in the soup. For example, one could easily assume that the ten proteins not found in Miller's table were also absent from the soup. With this single assumption, the information and molecular knowledge found in most proteins becomes infinite. Furthermore, the assumption to exclude chemicals like aldehydes and formic acid greatly improves the likelihood for protein evolution.

With these assumptions in place, labeling wooden blocks according to amino acid abundance yields table 5.1. The numbers in the second column are taken from Miller's table. The right column is based on what might have been given the constraints of the favorable assumptions discussed above.

Table 5.1: Wooden Blocks Used to represent Chemicals in the Soup

amino acid	number of blocks	amino acid	numbers of blocks
glycine*	440,000	tryptophan	400
alanine	395,000	tyrosine	400
valine	9,750	histidine	40
leucine	5,650	lysine	40
isoleucine	2,400	cysteine	400
proline	750	methionine	400
aspartate	17,000	phenylalanine	400
glutamate	3,850	arginine	40
serine	2500	asparagine	400
threonine	400	glutamine	400
Total number of blocks labeled with amino acids used by life		880220 (sum of column 2 and 4)	
Total number of blocks		4 x 880220 = 3,520,880	

* Most amino acids exist in two forms. The forms are mirror images of each other. Life only uses one image. Glycine is the only amino acid that does not have a mirror image. Thus, the number reported for glycine in table 5.1 corresponds to the concentration reported in Miller's table. The numbers associated with the other amino acids in this column are ½ the value reported in Miller's table.

The Evolution of a Functional Protein in the Primordial Soup

Because three million blocks cannot fit in a basket, the trapped scientist is now given a truck (figure 5.1). The blocks in the truck are determined by table 5.1. The scientist can draw blocks from a tube that connects his room to the back of the truck. How much information would insulin contain, if it evolves given these constraints.

Figure 5.1: Trapped Scientist with a Truck

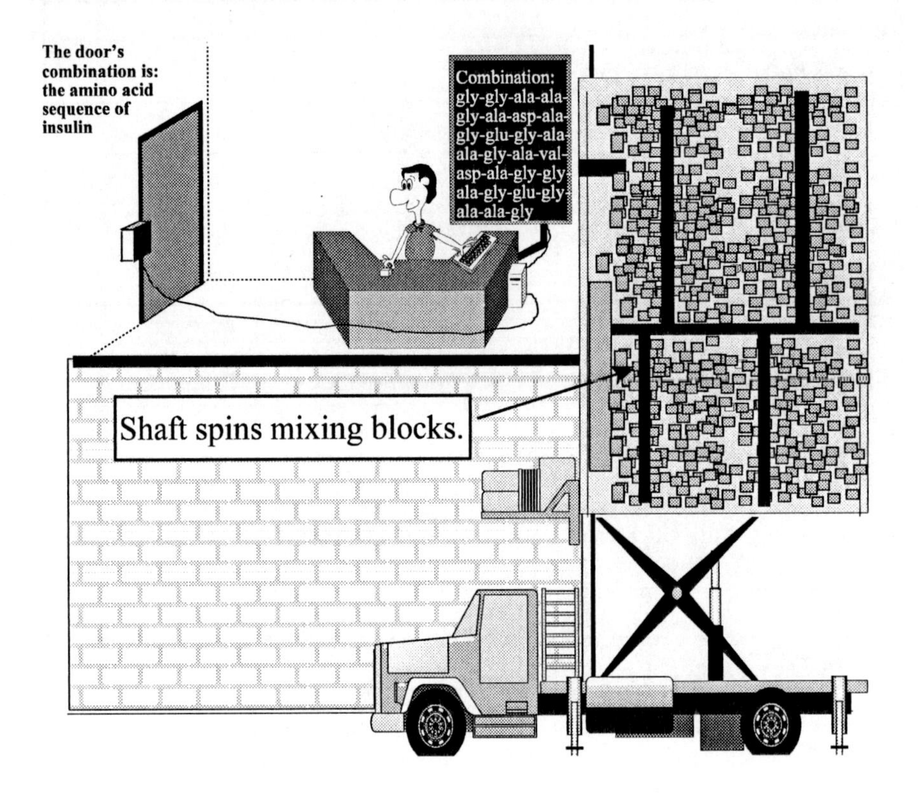

Table 5.2: Information in Insulin B Chain (Primordial Evolution)

pos	allowed amino acids	bits	pos	allowed amino acids	bits
2	phe, ala, leu, val	3.1	17	phe, tyr	12.1
3	val, ala, pro	3.1	18	leu	9.3
4	pro, lys, asn	11.5	19	val, ile	8.2
5	gln	13.1	20	cys	13.1
6	his, arg	15.4	21	gly	3.0
7	leu	9.3	22	asp, glu	7.4
8	cys	13.1	23	arg	16.4
9	gly	3.0	24	gly	3.0
10	ala, pro, ser	3.1	25	phe	13.1
11	his	16.4	26	phe, tyr	12.1
12	leu	9.3	27	tyr	13.1
13	val	8.5	28	thr, ser, asn	10.1
14	glu, asp	7.4	29	pro	12.2
15	ala	3.2	30	lys, arg	15.4
16	leu	9.3	31	ala, arg, thr, ser	3.1

Total bits = 280.5 bits

Example calculation: Phenylalanine, alanine, leucine and valine are possible at position one. There are 5,650 blocks labeled leu, 395,000 labeled alanine, 9,750 labeled valine, and 400 labeled phenylalanine in the truck. The total number of blocks is 3,520,880. So the probability that the scientist will pull a leucine, alanine, phenylalanine, or valine is (395,000+9,750+5,650+400) = 410,800 times in 3,520,880 tries. Thus the information at position one is calculated as follows: information = 3.32 x log (3,520,880/410,800) = 3.1 bits.

The total number of bits is 280.5. In chapter 4, the total for the B chain was only 108 bits. Intuitively, this is obvious because any proteins that emerge in the primordial soup will be composed of mostly alanine and glycine. Since real proteins do not follow this pattern, they are less likely to evolve in the primordial soup. The conclusion is that it is much harder for information and knowledge to evolve in the primordial soup.

The B chain of insulin contains 30 amino acids. So the average information contributed by each amino acid is equal to the total information divided by 30.

Information before life = 280.5/ 30 = 9.35 bits per amino acid

Information with the genetic code = 108/30 = 3.6 bits per amino acid

Because knowledge is defined in terms of information, it too must increase.

Molecular Knowledge in The Primordial Soup

Table 5.3 calculates the knowledge in the B chain of insulin assuming that the protein evolved in the primordial soup.

Table 5.3: Molecular Knowledge in Insulin B Chain

pos	allowed amino acids	bits	pos	allowed amino acids	bits
2	phe, ala, leu, val	2.0	17	phe, tyr ,(trp)	2.0
3	val, ala, pro	2.0	18	leu, (ile),(val), (ala), (met)	3.1
4	pro, lys, asn	2.0	19	val, ile, (ala), (leu), (met)	3.1
5	gln, (asn)	12.1	20	cys	13.1
6	his, arg, (lys)	14.8	21	gly	3.0
7	leu, (ile), (leu), (val), (met)	3.1	22	asp, glu	7.4
8	cys	13.1	23	arg, (lys), (his)	14.8
9	gly	3.0	24	gly	3.0
10	ala, pro, ser	2.0	25	phe, (tyr), (trp)	11.5
11	his, (lys), (arg)	14.8	26	phe, tyr, (trp)	11.5
12	leu, (ile), (val), (ala), (met)	3.1	27	tyr, (phe), (trp)	11.5
13	val, (ile), (leu), (ala), (met)	3.1	28	thr, ser, asn	2.0
14	glu, asp	7.4	29	pro	12.2
15	ala, (leu), (ile),(val), (met)	3.1	30	lys, arg, (his)	14.8
16	leu, (ala), (val),(Ile), (met)	3.1	31	ala, thr, ser, -	0*

Total = 211 bits.

* Any position with a gap does not need an amino acid and therefore the knowledge is set to 0 bits.

Example calculation: At position 7, only leucine is found in the alignment. Nevertheless, the technique to calculate knowledge assumes that the other amino acids in this group are allowed. The number of blocks labeled with the 5 amino acids belonging to group 1 in the truck (figure 5.1) is 413,200. There are 3,520,880 total blocks. So the knowledge is 3.32 x log (3,520,880/413,200) = 3.1 bits.

Notice that no position can ever contribute less than 2 bits. If all 20 amino acids are found at a particular position, the position still contributes 2 bits. This accounts for the amino acids not used by life found in the soup.

The average knowledge per amino acid in the soup is calculated as follows: knowledge = 211 total bits / 30 amino acids = 7 bits per amino acid. The average knowledge per amino acid with the genetic code is only 76 total bits /30 amino acids = 2.5 bits per amino acid. (Refer to pg. 76, table 4.4 for number of bits using the code).

Because of the nature of logarithms, the implications are dramatic. Suppose that one of the first proteins to evolve contains 100 amino acids, and that 30% of this protein shows a conservation pattern similar to insulin.

Knowledge today = molecular knowledge =
100 amino acid x 2.5 bits per amino acid x 30% = 75 bits

Odds of evolving today are 1 time in 2^{75} tries or 1 time in 4×10^{22} tries. This could happen with enough tries.

Knowledge soup = molecular knowledge + primodial information

Knowledge soup = 100 x 7 bits per amino acid x30%
+ 100 x 2 bits per amino acid x 70%
350 bits

Odds of evolving in the primordial soup are 1 time in 2^{350} tries or 1 time in 2.2×10^{105} tries. This can never happen.

Before life exists, chance will require an incredible number of tries to create knowledge, and the vastness of space, the number of atoms in the universe, and the incredible age of the universe do not make a dent in the problem. Nature simply cannot accumulate enough tries to overcome the poor odds.

Finally, this chapter had to make quite a few assumptions. Some readers may be concerned about these assumptions, but realize that almost every assumption was for the benefit of protein evolution. For example, this chapter assumed that the primordial soup did not contain aldehydes, carboxylic acids, and amines. This assumption is obviously false, but it greatly improves the chance for a protein evolving because it eliminates many side reactions. Also the amino acids not listed in Miller's table because they are not present in significant quantities are assumed to be in the soup at a very generous level. Allowing every star in the universe to have one planet is certainly a generous assumption, but perhaps the most generous assumption is to allow every single one of these planets to be composed entirely of peptide chains each containing 150 amino acids. This assumption is only rivaled by the next one that allows these peptide chains to break down and reform every year, and even with all of these generous assumptions, the probabilities do not budge from zero.

Any scientist who believes that nature can create molecular knowledge before life exists is relying on faith to justify his opinion because the math just does not support this belief.

References:

1) Keefe and Szostak, "Functional Proteins from a Random Sequence Library," Letters to Nature, 410: 715-718, 2000.

2) Miller, Which Organic Compounds Could have Occurred on the Prebiotic Earth?, Cold Spring Harbor Symposium of Quantitative Biology Volume L11, 17-25, 1987.

3) Miller, Orgel, The Origins of Life on Earth, Prentice Hall, 1974

4) Ehrenfreund, Glavin, Botta, Cooper, Bada, "Extraterrestrial amino acids in Orguil and Ivuna: Tracing the parent body of CI type carbonaceous chondrites," PNAS, 98: 2138-2141, 2001.

5) Fox, Dose, Molecular Evolution and the Origin of Life, 1972.

Chapter 6: Introduction to Chemistry and Entropy

This chapter will introduce chemistry, organic chemistry, quantum mechanics, and thermodynamics. The goal is to make sure that all readers understand how and why chemicals react with each other, and how and why the laws of thermodynamics influence these reactions.

The key concept of entropy will be introduced in this chapter. Entropy is often defined as disorder, but this definition is both misleading and incorrect. In classical thermodynamics, entropy is a mysterious concept. Entropy is difficult to define without considering quantum mechanics and micro-states. While these last two topics are usually only found in advanced chemistry and physics textbooks, they are absolutely necessary to understand entropy. Entropy is not a difficult concept. It is simply a measure of uncertainty that must always increase with time.

Entropy makes it very difficult for a self replicating molecule to exist because self replication decreases entropy. Life has many ways around this problem. The most common solution involves tapping plentiful energy sources to drive replication (chapter 7). Simple self replicating molecules cannot do this.

Chemicals and Atoms

Chemicals make up everything that is a solid, liquid or gas. The earth's atmosphere today is composed predominantly of two chemicals, nitrogen and oxygen. Chemicals are made from atoms. A chemical composed entirely of the same type of atom is called an element. Oxygen is an element because it is composed entirely of oxygen atoms. Hydrogen is an element because it is composed of hydrogen atoms. Water is not an element because it is made from oxygen and hydrogen atoms. A molecule is a collection of two or more atoms. A water molecule consists of two hydrogen atoms and one oxygen atom.

The Hydrogen Atom

Hydrogen is the simplest atom. It contains one proton and one electron. The proton is the central core, the nucleus. Contrary to popular belief, electrons do not orbit the nucleus like the planets orbit the sun. When a small particle like an electron is confined to a small space, it no longer behaves like a particle. Its energy now must jump in discrete steps. This situation is analogous to a baseball that can only be thrown at 20, 50 or 90 mph. If a pitcher tries to throw it at any other speed, the baseball will still travel at 20, 50 or 90 mph. Furthermore, the electron can never be precisely located at any instant in time. The electron is always found somewhere in its orbital (figure 6.0).

When two electron orbitals merge to form a hydrogen molecule, the two electrons are free to occupy either orbital. Both hydrogen atoms share these two electrons. Because each electron can be found anywhere in this merged orbital, the electrons can find a lower energy state. This is why chemical bonds form.

Figure 6.0: Hydrogen Atom

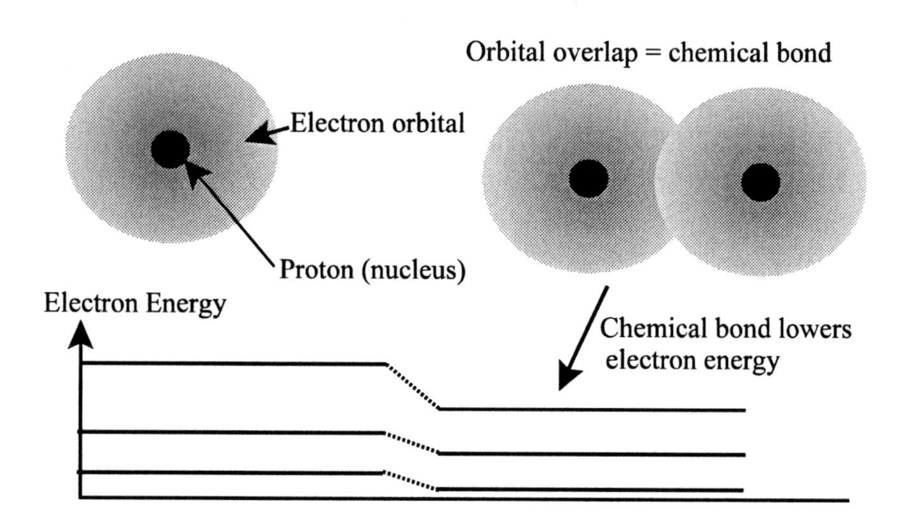

Representing Chemicals with Symbols

Every atom can be represented by a symbol. These symbols are usually the first letter in the name. For example, the symbol for hydrogen is H, the symbol for carbon is C, the symbol for oxygen is O, the symbol for nitrogen is N, and the symbol for sulfur is S.

Using symbols, a molecule of water is represented by H_2O where the subscript indicates that there are two hydrogen atoms in every water molecule. Figures 6.1 and 6.2 show how chemicals can be represented by names, symbols, balls and sticks, and spheres. The lines connecting the atoms are chemical bonds. Chemical bonds are the glue that hold molecules together.

In figure 6.1, three chemicals are represented, oxygen, water and methane. The chemical formulas for each are O_2, H_2O and CH_4. The ball and stick models make it easy to see what chemical looks like, but the ball and stick models do not show the electron orbitals. Atoms are really much bigger as shown in the space filling models of figure 6.1.

This book will also generate many images using a molecular visualization tool called Rasmol. The images of chemicals are more accurate with this tool. Figure 6.2 shows the amino acid, valine. Because the atoms are no longer labeled, the color must be used as an indication of atom type. In figure 6.2, black is oxygen, white is hydrogen, light gray is carbon, and dark gray is nitrogen. Rasmol can generate a space fill image, a ball and stick image, and a stick image as shown in figure 6.2. Rasmol also allows the image to be rotated by the user. While the space fill model in figure 6.2 is the only accurate representation of valine, it is too difficult to see how the atoms are connected using this model. So the stick and ball and stick are often preferred. In complex molecules, the cartoon view generated by Rasmol is often preferable (figures 3.13 and 3.14 on pages 58 and 59).

Figure 6.1: Ball and Stick and Space Filling Model

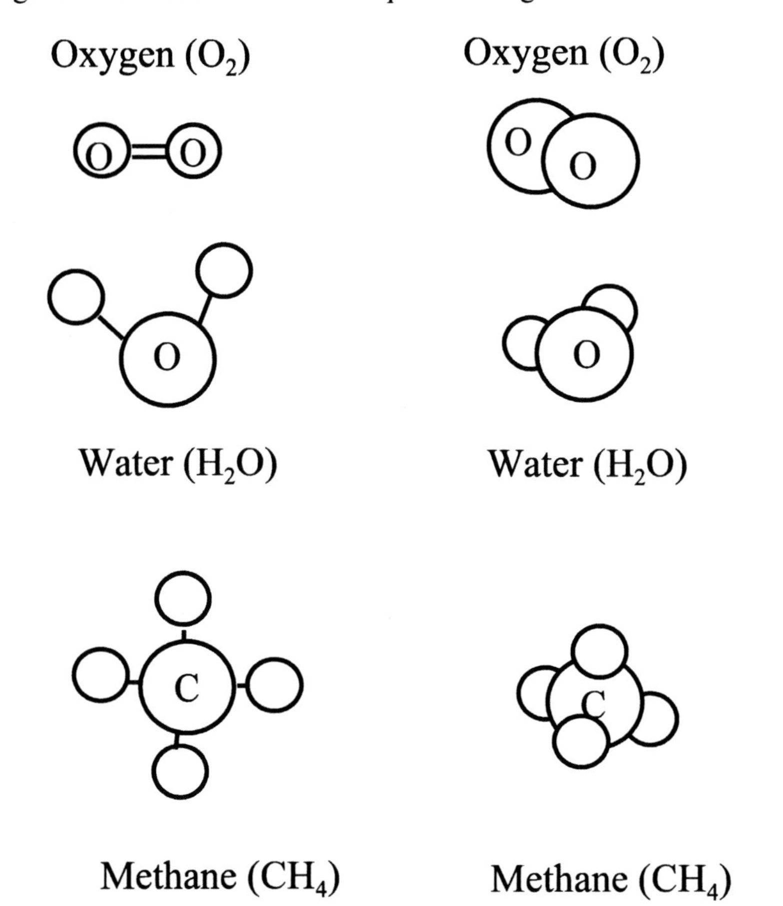

In both models, unlabeled small circles are hydrogen.

Figure 6.2: Rasmol Images of Valine

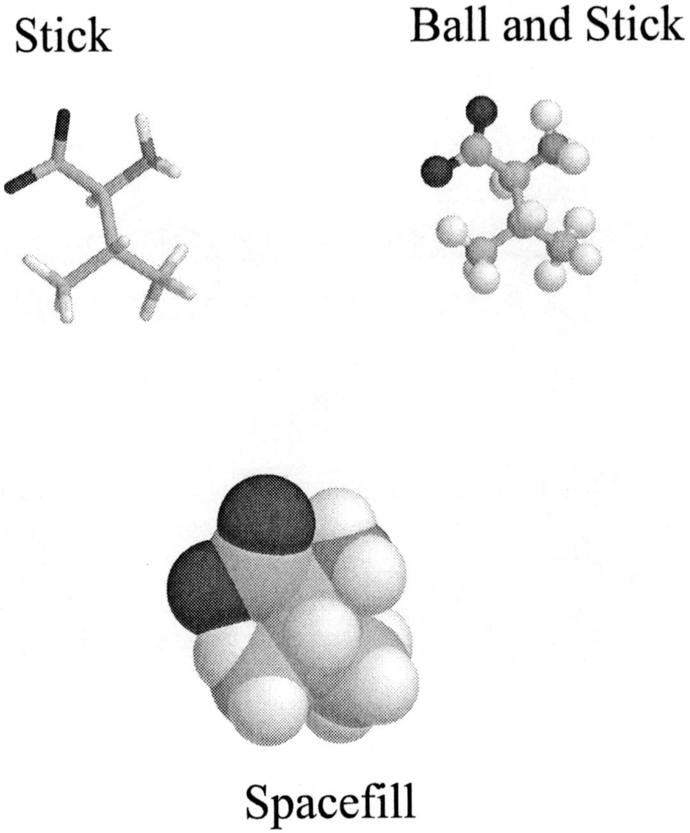

Stick

Ball and Stick

Spacefill

Chemical Bonds

Atoms are composed of three components, electrons, neutrons, and protons. The protons and neutrons form the central core, the nucleus, and electron orbitals interact with each other to form chemical bonds in such a way that each orbital is filled with electrons. In many chemicals, this can only be accomplished by sharing the available electrons.

To fill their orbitals, carbon atoms must share 4 electrons. To do this, they need to form 4 chemical bonds. Nitrogen atoms must share 3 electrons, so they typically form 3 chemical bonds. Hydrogen atoms must share 1 electron, so they typically form one chemical bond. Oxygen atoms must share 2 electrons, so they typically from two bonds.

By knowing how many chemical bonds each atom requires, it is possible to predict chemical structures. For example, every water molecule has 2 hydrogen atoms and 1 oxygen atom. The oxygen atom is satisfied because it shares 2 electrons, one with each hydrogen atom. Likewise, both hydrogen atoms are satisfied because each shares one electron with the single oxygen atom (figure 6.1).

Multiple Bonds

In many cases, if atoms cannot satisfy their requirements for sharing electrons, they will form double and triple bonds. For example, oxygen forms a double bond with itself. Since oxygen desires two electrons, by forming a double bond instead of a single, each oxygen atom in an oxygen molecule is satisfied (both have 2 bonds and share 2 electrons).

Chemical Symbols

It is not practical to draw the ball and stick model or the space filling model every time a chemical is mentioned. So instead chemists have developed various shorthand representations. The most condensed is to simply use symbols to represent chemicals. The symbol H_2O represents a water molecule.

This shorthand is not that useful for large molecules because it does not indicate how the various atoms are arranged. It is easier to visualize chemicals with the ball and stick model, but these are too cumbersome to draw. So the compromise is to replace the ball in the ball in stick model with a letter representing the atom. So hydrogen atoms are replaced by the letter H, and oxygen atoms are replaced by the letter O. Lines then connect the symbols to show chemical bonds. Since carbon often forms the backbone of large molecules, it is generally depicted by a line that bends. The hydrogen atoms connected to this assumed carbon are also assumed. This technique is shown for alanine in figure 6.3.

Figure 6.3: Alanine

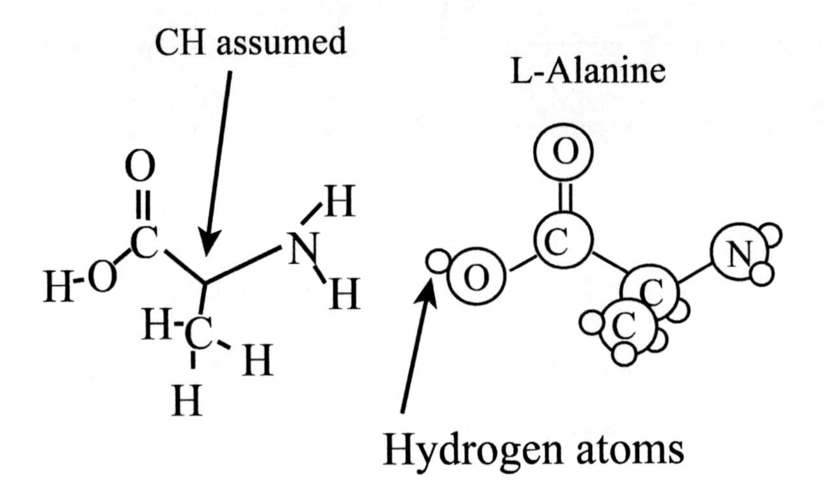

Matter, Energy, Heat, and Temperature

Matter is composed of atoms. Matter takes three forms, solid, liquid and gas.

Temperature is a measure of how fast the atoms in matter are moving. If a room is hot, then the oxygen and nitrogen atoms in the room are moving very fast. If the room is cold, the atoms are moving slowly. The same is true in water. Even in a solid rock, the atoms are free to vibrate, and this vibration is a measure of the temperature.

Heat is a measure of energy transfer. Heat always flows from hot objects to cold ones. Fast moving atoms impart some of their energy to slower ones when they collide. This slows down the fast atoms and speeds up the slow ones. Energy is thus transferred from hot objects to cold ones.

Energy is the ability to do work. A boulder sitting on top of a hill is said to have potential energy. When it starts rolling down the hill, this potential energy is converted to kinetic energy. If this boulder is tied to a rope attached to a cart on the other side of the hill, then it can lift the cart up the other side as it rolls. Lifting the cart is work. So in this example, the boulder accomplishes work as it rolls.

As the boulder rolls some of its potential energy will be converted to heat. This will raise the temperature of the boulder, the hill, the air, and the cart. This means that some of the work done in lifting the boulder to the top of the hill cannot be recovered when the boulder rolls down the hill.

Quantum Mechanics

As mentioned earlier when a small particle like an electron is confined to a small space it no longer behaves like a particle. Its energy becomes quantized. This means that it can only take on discrete energy levels. The best way to illustrate this is to envision a boulder rolling down a hillside. In classical physics, the hillside is gently sloped allowing the boulder to be anywhere on the hill (figure 6.4). It gradually gains speed as it progresses down the hill. The loss in potential energy is continuous. Some of this energy becomes kinetic energy (the energy associated with the moving boulder) and some of it becomes heat. Quantum physics tells a different story. The hill is like a series of steep cliffs separated by flat areas. Atoms and electrons can only reside on the plateaus. Each plateau is a quantum energy level. For atoms and molecules, this quantum energy level is similar to kinetic energy. That is faster moving atoms occupy a higher energy state. Atoms can rotate, vibrate, and move through space. All of these energy states are quantized.

Figure 6.4: Quantum Physics vs. Classical Physics

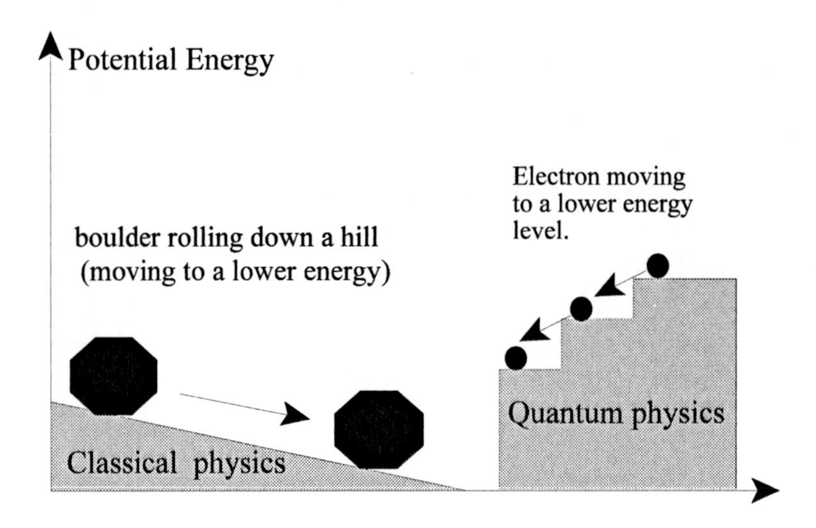

Micro-states and Entropy

Consider two very cold chambers that are completely empty except for a few oxygen atoms (figure 6.5 and 6.6).

Chamber 1: 4 atoms, 5 quantum states, total energy = 8 units
Chamber 2: 3 atoms, 5 quantum states, total energy =6 units

There are 5 quantum energy states available to these atoms. The lowest has an energy of 2 units, and the highest has an energy level of 10 units. The atoms in both chambers are only free to occupy the lowest energy state. If one occupies a higher energy state then the total energy of the system will be too high. In this case, there is only one possible way to arrange the atoms in each chamber. The ways that the atoms are arranged to fill the quantum energy states are called micro-states. Figures 6.5 and 6.6 illustrate this principle.

Figure 6.5: Micro-state for Chamber 1, Total Energy = 8 Units

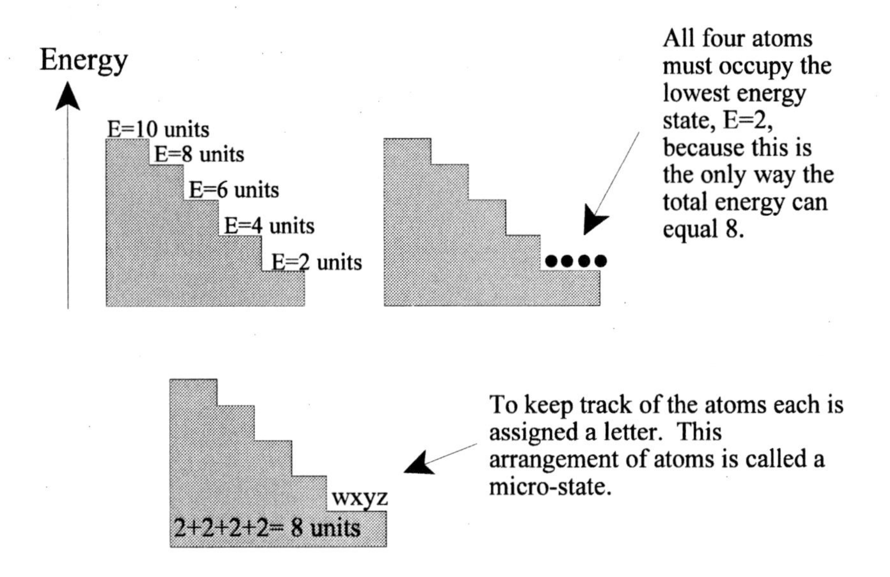

Figure 6.6: Micro-states for Chamber 2, Total Energy = 6 Units

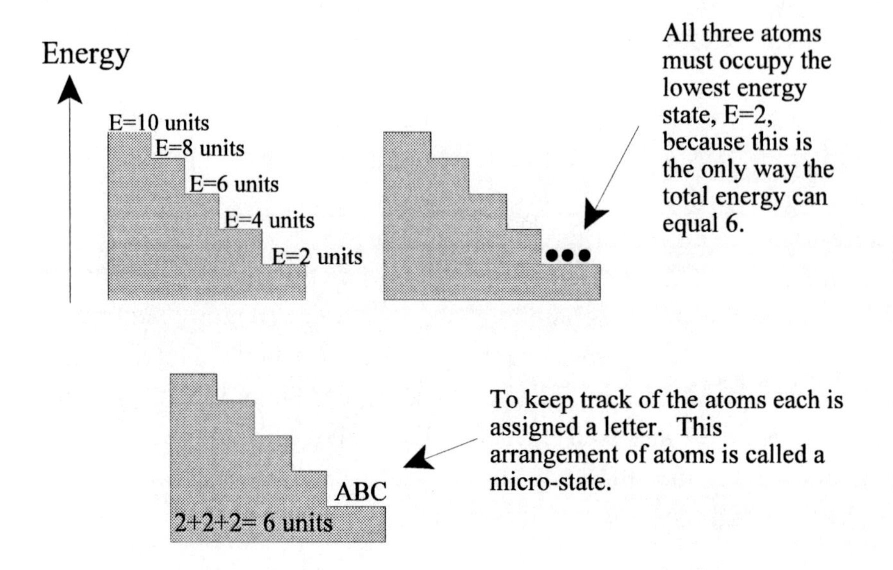

If chamber 2 is heated, the atoms will move faster. Assume the total energy is now 12 units. Figure 6.7 shows that the number of available micro-states is greatly increased.

Figure 6.7: Ten Micro-states Available After Heating

Distribution 1: all atoms have 4 energy units

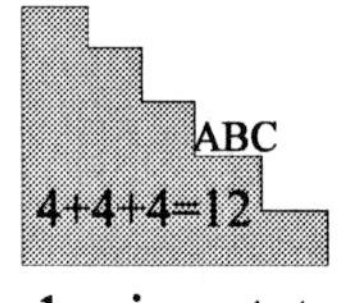

1 microstate

Distribution 2: one atom has 8 units and two have 2 units

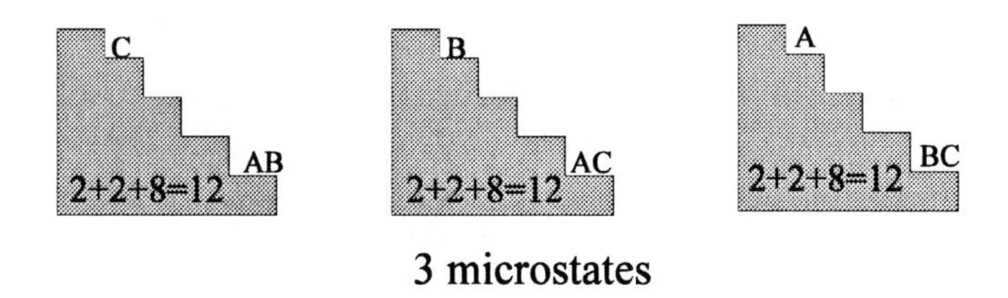

3 microstates

Distribution 3: one atom has 6 units, one has 4 and one has 2

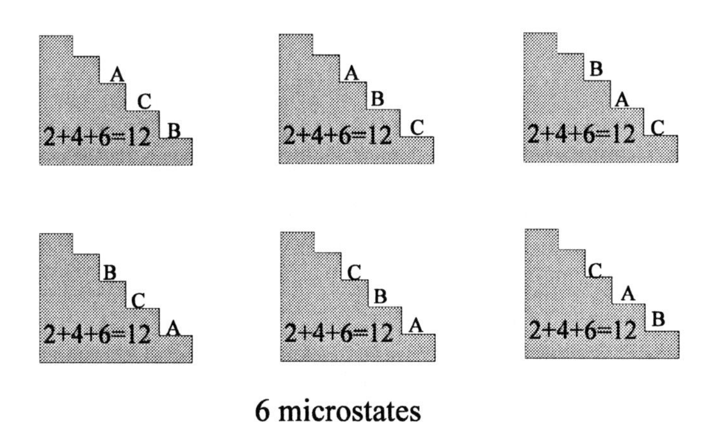

6 microstates

Entropy is often defined as a measure of disorder, but this definition is not only misleading it is wrong. Entropy is a measure of available micro-states. So in this example, the oxygen atoms that are heated have more entropy (10 micro-states available vs. 1 micro-state available).

Entropy can also be defined as a measure of uncertainty. Because as more micro-states become available to the system, the state of the particles becomes more uncertain. Entropy has nothing to do with disorder.

From this example, it should be clear that there are several ways to increase the entropy of a system. Increasing the temperature is one. Suppose 5 quantum energy levels are added so that all of the odd energies are represented, E=1, E=3, E=5, E=7, and E=9. The oxygen atoms will be able to distribute themselves in many more ways and thus find more micro-states.

The Second Law

The second law of thermodynamics states that in any spontaneous process the entropy of the universe will increase. What does this mean? It means that all spontaneous processes must increase the number of available micro-states. The number of available micro-states after any event will always be greater than the number of available micro-states before the event. Because atoms form large objects like boulders, large objects must also obey the second law.

The second law can be stated in a very intuitive way. The uncertainty of the universe increases with time. This is why it is more difficult to predict what will happen far in the future. Weather is a great example. The weatherman may be able to forecast rain tomorrow, but he cannot forecast rain a month in advance.

Consider a bicycle that is turned upside down. The back wheel is spun until it is moving very fast. The second law explains why the wheel will not turn for long. The atoms that make up the wheel are moving very fast, so these atoms have lots of energy. As the wheel spins some of this energy is transferred as heat to the air around the wheel and to the frame that holds the wheel. This increases the air temperature which in turn increases the number of micro-states available to the air molecules. The entropy of the air molecules increases. Since the frame also heats up, its atoms are free to occupy more micro-states. Eventually all of the energy in the wheel will be dissipated as heat. The wheel's entropy decreases as it slows. The entropy of its surroundings increase. The increase is more than enough to offset the decrease. Thus, the entropy of the universe as a whole increases.

Figure 6.8: System vs. Surroundings

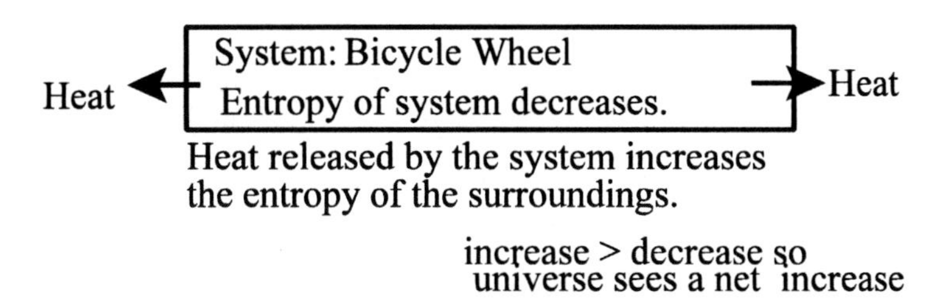

The equation that must always be satisfied is as follows:

entropy change of the system + entropy change of the surroundings = entropy change of the universe > 0.

The entropy of any system can decrease as long as the entropy of the surroundings increases, and the increase is greater than the decrease.

Heat Flows From Hot Objects to Cold Ones

Suppose that chamber 1 (figure 6.5) and chamber 2 (figure 6.7) are brought together so that their walls touch. Heat can now be transferred between the systems, but the atoms are confined to their respective chambers. Because chamber 2 is hot and chamber 1 is cold, heat should flow out of chamber 2 and into chamber 1. Does such a flow increase the number of available micro-states? Figure 6.9 and 6.10 show that this process increases the number of available micro-states. The steady state (maximum number of micro-states) is reached when energy is distributed equally among both chambers.

This example in meant to convey an intuitive feel for the second law, what it means, and how it works. Keeping track of how micro-states change in real processes is not practical. There are too many atoms and too many quantum states. The number of available micro-states in most systems is far greater than the number stars in the universe. Even the most powerful computer cannot keep track of this many micro-states.

Fortunately, when the number of atoms is increased to 10,000 or more, one distribution dominates. In figure 6.7, distribution three is the most probable. The system will spend 60% of its time in one of the micro-states belonging to this distribution. As the number of atoms is increased, the dominance of the most probable distribution also increases. With 100,000 or more atoms, the system will spend all of its time in the most probable distribution because the most probable distribution is always the one with the most available micro-states.

Entropy is now easy to understand. As chemicals or atoms react and move around, they try to find their most probable distribution. Since this is the distribution that maximizes the number of available micro-states, it is also the distribution that maximizes entropy. Thus, the entropy of the universe always increases.

Figure 6.9: Initial Distribution of Micro-states

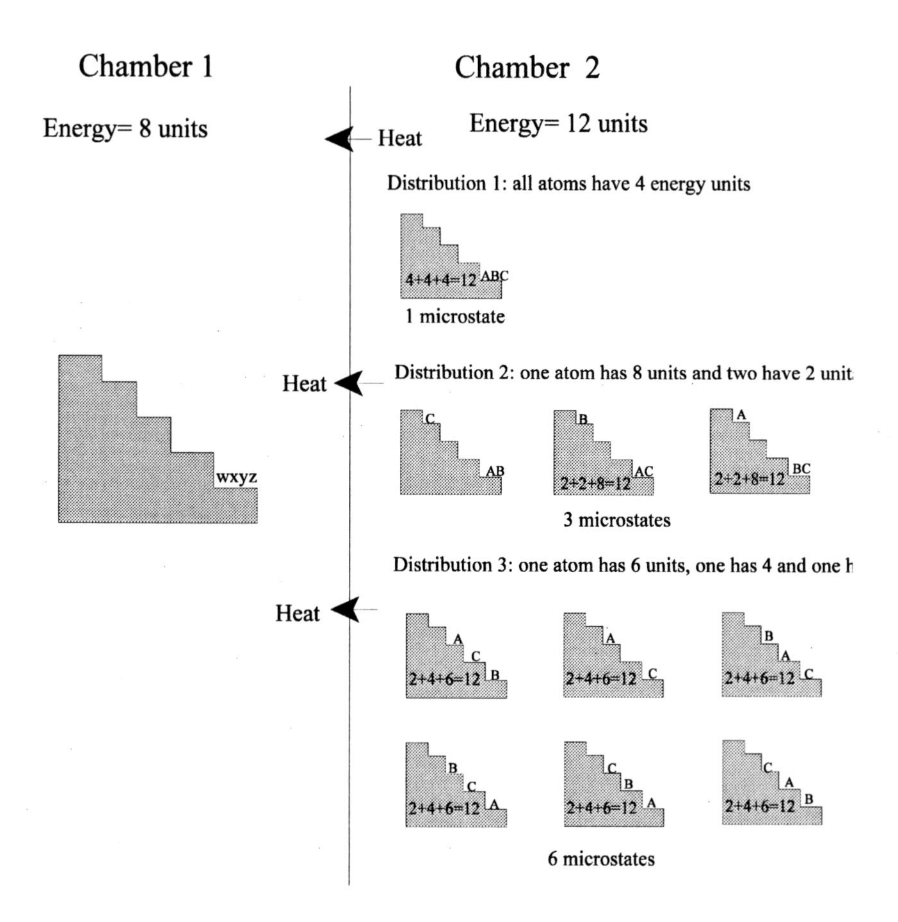

Mathematically, if the universe is defined as both chambers, then there are only 10 micro-states (not 11). No matter which micro-state chamber 2 chooses, chamber one is always in the same micro-state. Thus the total number of micro-states is given by multiplication not addition and 10 x1 =10.

Figure 6.10: Final Distribution of Micro-states

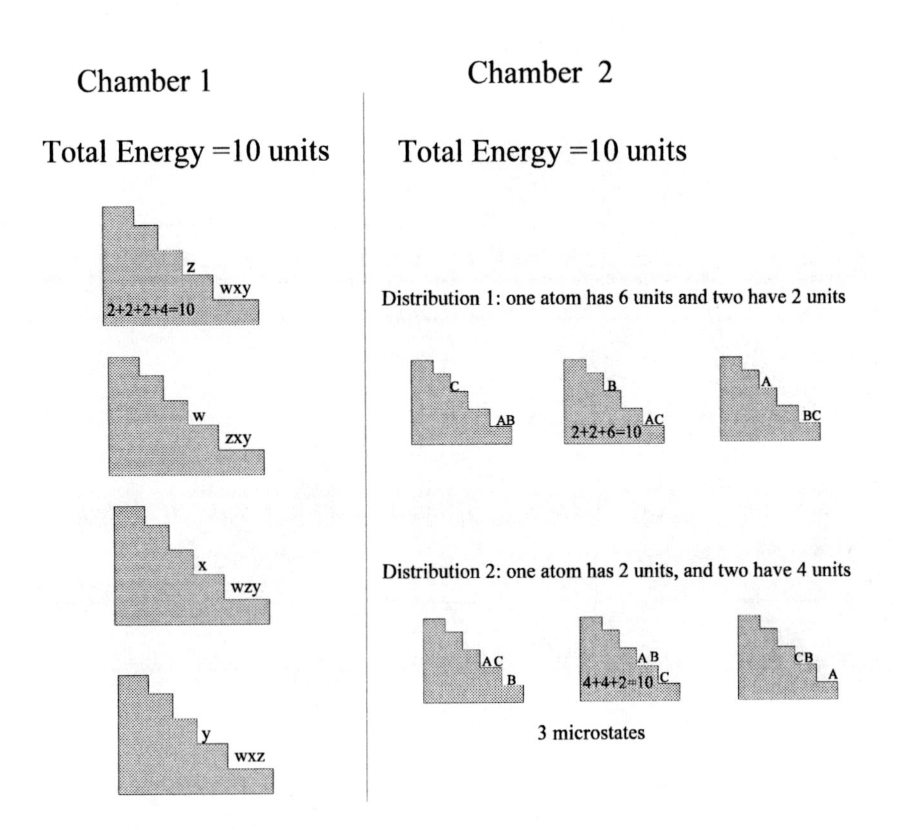

Mathematically, if the universe is defined as both chambers, then there are only 24 micro-states (not 10). For each micro-state chamber 2 chooses, chamber one can choose 4. Thus the total number of available micro-states is given by multiplication not addition and 4x6 =24. Thus, when heat flows from a hot object to a cold one, the number of microstates and hence the entropy of the universe is increased.

Entropy and Chemical Reactions

Chemicals interact with each other in chemical reactions. Chemical reactions break existing chemical bonds and form new ones. The reaction is represented symbolically by an arrow. Consider the following chemical reaction: $2H_2 + O_2 \rightarrow 2H_2O$. In this reaction two molecules of hydrogen combine with one molecule of oxygen to yield two molecules of water (figure 6.11).

The second law determines whether or not this reaction will happen. Let the chemicals be the system. The electrons in water have less energy than those in hydrogen and oxygen. This decrease in electron energy releases heat as the reaction takes place. This heat increases the entropy of the surroundings, figure 6.12. This heat drives the reaction forward because it ensures that the entropy of the universe increases. Thus, this chemical reaction is said to be spontaneous. This means that given time, the reaction will happen. It does not mean that the reaction will happen quickly.

Figure 6.11: A Simple Chemical Reaction

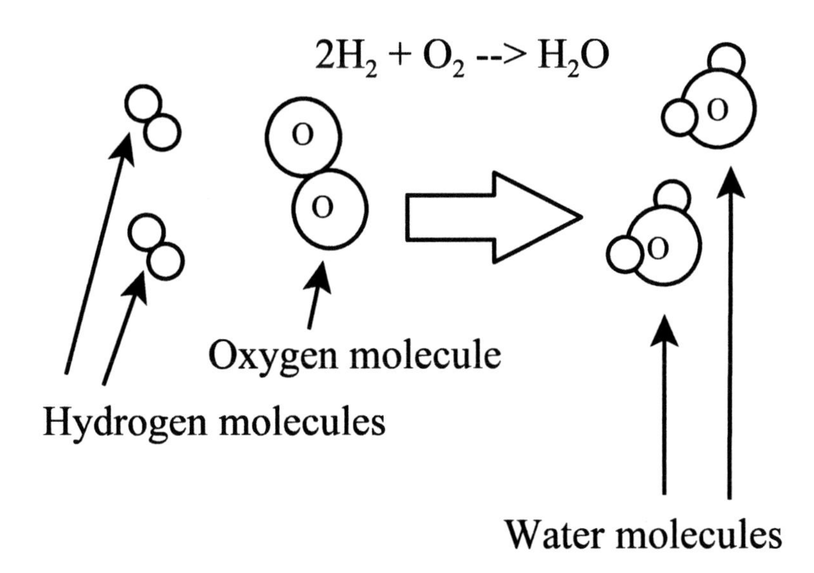

Figure 6.12: System Diagram for Above Reaction

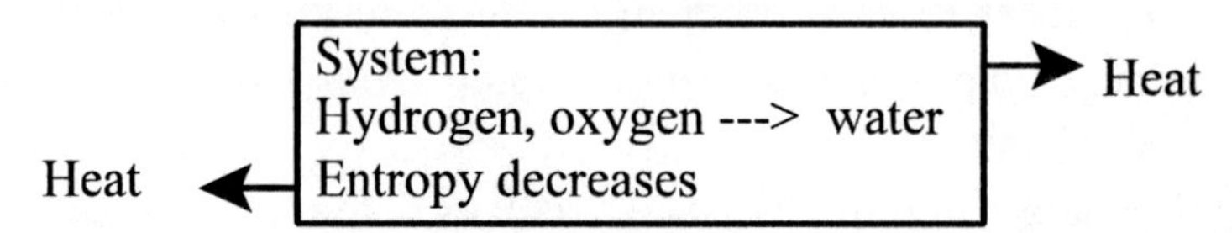

Heat released by system increases the entropy of the surroundings

increases > decrease so the universe realizes a net increase

Notice that the entropy decreases for the isolated system because when the system exists as oxygen and hydrogen more micro-states are available than when the system exists as water, but just like the bicycle example, the heat released increases the entropy of the surroundings. This allows the entropy of the universe to increase. Therefore, the reaction is spontaneous.

Chemical Kinetics

When hydrogen and oxygen are mixed together in a chamber at room temperature, nothing happens. There is no chemical reaction, but if a match is lit in the chamber, the chemical reaction happens so fast that it is explosive.

The reaction of hydrogen and oxygen to form water is spontaneous, but it will not happen unless some energy is put into the system. The energy required is called the activation energy. The chemical equation

$2H_2 + O_2 \rightarrow 2H_2O$ can be written as

$2H_2 + O_2 \rightarrow$ very high energy intermediate state $\rightarrow 2H_2O$

to indicate that for the reaction to happen the chemicals involved must transition through a short lived high energy intermediate state.

Because this state has more energy than the initial and final states, energy must be put into the chemicals to allow them to reach the intermediate state. In this case, the match allows some hydrogen and oxygen atoms to transition to the intermediate state. They then form water and release heat. The heat release causes more hydrogen and oxygen atoms to transition to the intermediate state. The reaction builds on itself in this manner until almost all of the hydrogen and oxygen are used up and only water remains.

Activation energy is very important for life. It allows chemicals to exist in a state indefinitely even if a change in state may increase the entropy of the universe. Figure 6.13 conveys these key concepts. The reaction to create water will increase the entropy of the universe as indicated by the arrow on the left side of the figure. Therefore, the reaction is said to be spontaneous. Nevertheless, it is not spontaneous unless the hydrogen and oxygen are provided with enough energy to cross through the high energy intermediate stage.

It is the activation energy that allows life to exist. Because of this barrier, chemicals that are not thermodynamically favored can exist for many hundreds of years. If a process increases the entropy of the universe, then the second law defines the process as spontaneous, but it does not have to happen right away. The process may take years to complete. The speed of a chemical reaction depends on the activation energy. The second law does not determine how fast a chemical reaction will happen. Hydrogen and oxygen can coexist in a chamber for a thousand years if no energy source is present to start the reaction.

Notice that the direction of increasing entropy is drawn downward in figure 6.13 to indicate that water is the preferred state as it maximizes the entropy of the universe.

Figure 6.13: Activation Energy

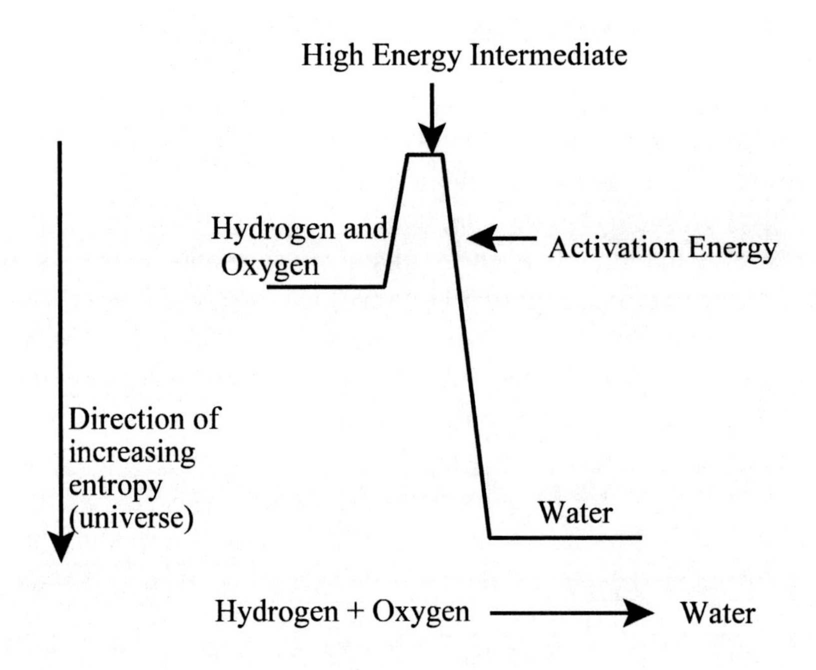

Chemical Equilibria

The reaction of hydrogen and oxygen to form water only happens in one direction. Figure 6.13 illustrates why. It is almost impossible for water to cross the activation barrier. Many chemical reactions happen in both directions. Figure 6.14 shows the reaction of a glycine-glycine molecule with water to yield two glycine molecules. Figure 6.15 shows how this reaction affects entropy. Notice that the entropy change is very small. The small change in entropy means that the reaction happens in both directions.

Figure 6.14: The formation of a glycine-glycine

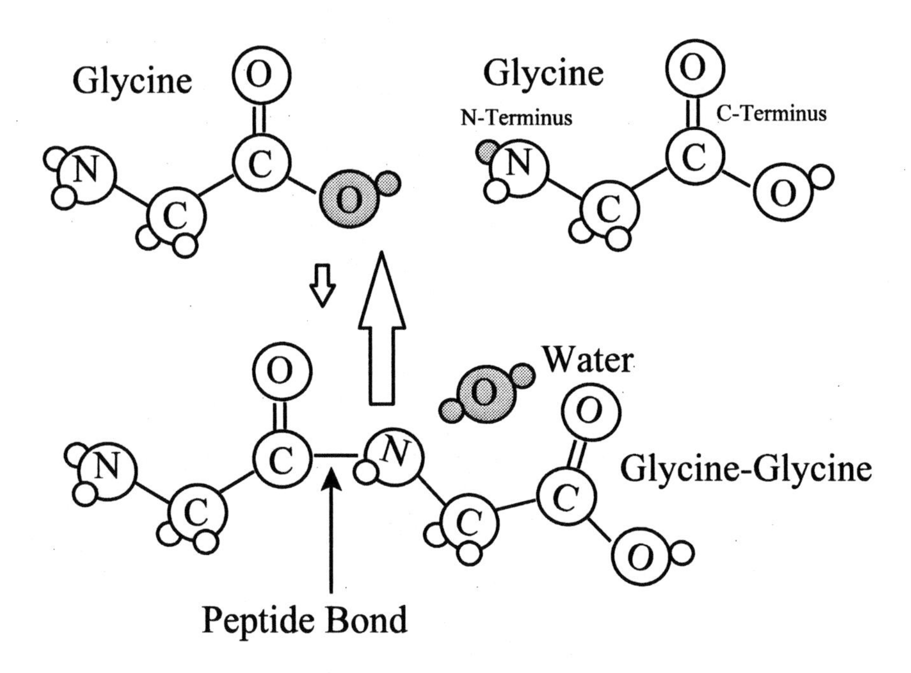

Figure 6.15: Entropy Change Associated with a Peptide Bond

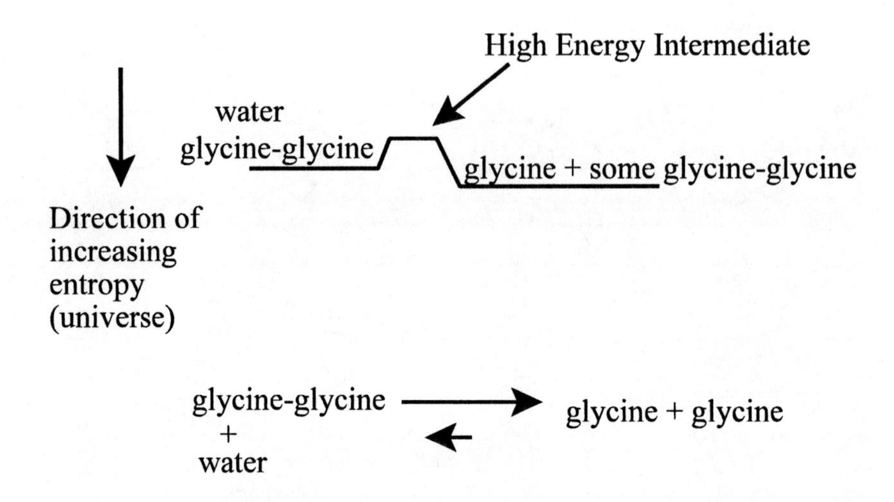

In this case, the energy barrier favors the formation of glycine + glycine, but the reverse direction also happens as indicated by the reverse arrow in figure 6.14 and 6.15. The number of available micro-states in this reaction is maximized when there is some glycine-glycine and quite a bit of free glycine. To satisfy the second law, this chemical reaction will find the point that maximizes the available micro-states.

The optimal mixture is the one that maximizes the available micro-states and hence the entropy. Figure 6.15 should be compared to figure 6.13. The change in entropy is so great in figure 6.13, that the number of available micro-states is maximized when the universe exists only as water.

Notice that this chapter uses the term *available micro-states* as opposed to *micro-states*. The number of micro-states is a property of a system and its surroundings, and as such, in many reactions the total number of micro-states does not change, but as unavailable micro-states become accessible to more atoms and electrons, the number of available micro-states increases.

At equilibrium the concentration of the chemicals in a system no longer changes. That is in figure 6.15, the concentration of glycine and glycine-glycine remains constant once the system reaches chemical equilibrium. The forward and reverse reactions still take place, but they cancel each other. Thus, no net change is observed.

Closed vs. Open Systems

Chemical equilibrium is a hard state to maintain. Almost anything that changes will alter equilibrium. If the temperature rises or falls, a new equilibrium will have to be found. If energy is put into the system, the temperature will rise. Any chemicals that enter or leave the system will also change equilibrium. Because of these factors, only closed systems reach equilibrium. A closed system is one that is completely isolated from its surroundings. Chemicals in the system are not allowed to leave. New chemicals are not allowed to enter, and no heat can be transferred to or absorbed from the surroundings.

The earth is an open system. The sun continually transfers energy into the earth's system, and the amount of energy varies with the time of year. On a smaller scale, the earth's oceans are open systems. They continually receive new water and chemicals from rivers, and lose water to evaporation. Lakes and ponds are also open systems. Closed systems are very rare.

The implication is that most chemicals do not ever reach chemical equilibrium. This is why the second law is often stated as follows: all spontaneous processes tend to increase the entropy of the universe. The second law does not state that all spontaneous processes must instantly maximize the entropy of the universe. This is very fortunate for life. It allows the chemicals in life to exist in a state very far from equilibrium. This is why life is possible. This topic will be discussed in the next chapter.

References:

Vemulapalli, Physical Chemistry, Prentice Hall 1993

Energy Flow in Biology, Morowitz, Ox Bow Press, 1979.

Chapter 7: Implications of The Second Law

This chapter will investigate the effects of the second law on life, and then show why the law makes it so difficult to explain the origin of life. Finally, the claim made by some authors that the second law makes evolution after life exists impossible will be investigated.

How Does Life Exist So Far From Equilibrium?

The answer lies with proteins. A specific kind of protein called an enzyme lowers the activation energy of chemical reactions. Reactions that would normally take years happen in seconds. Figure 7.1 shows how this effect alters chemical equilibrium. Consider a set of chemicals that have two paths to interact as shown in figure 7.1. One reaction is a side reaction that is undesirable. The other is desirable.

The enzyme lowers the activation energy for the desirable reaction making it happen quickly. The undesirable reaction does not have a chance. Also notice that the entropy of the universe is maximized by the undesirable reaction. Thus, from thermodynamic considerations, one might think that the undesirable reaction is always dominant, but because the laws of thermodynamics have no way to deal with time, this observation is seldom true. The enzyme is not violating the second law by forcing the reaction in the preferred direction. Its reaction is also spontaneous because it also increases the entropy of the universe. By making the desired reaction happen faster, the enzyme does not give the undesirable reaction time to happen.

Figure 7.1: Enzymes and the Second Law

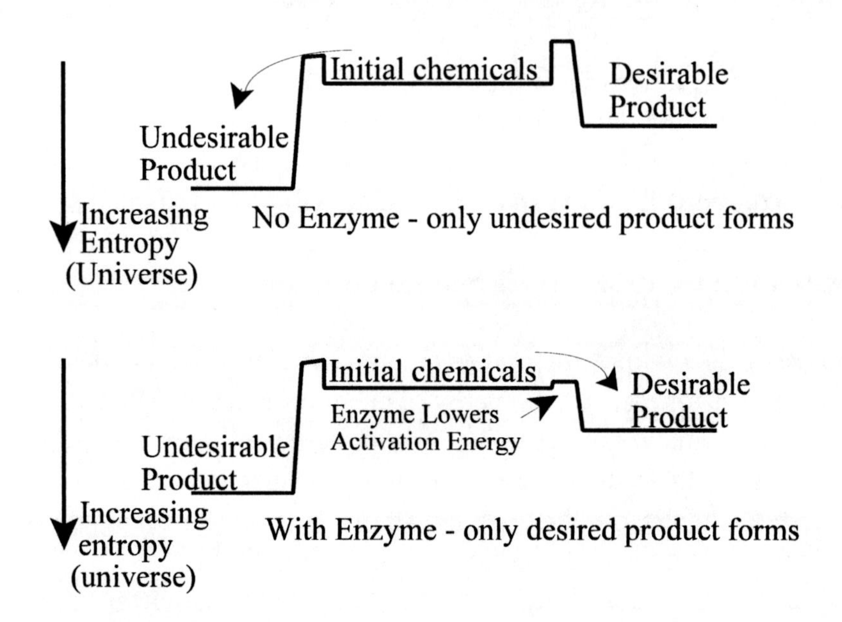

Figure 7.1 illustrates a powerful technique, but this trick alone does not enable life. Enzymes have another much more impressive trick. They can force chemicals reactions that should never take place to happen and happen quickly. They do this by coupling a favorable chemical reaction to an unfavorable one.

Lowering the activation energy will not help a reaction go forward if the entropy of the final state is lower than the initial. In figure 7.1, the final state of the desired products is a higher entropy state than the initial state (in these diagrams moving down hill represents increasing entropy). Thus, the chemical reaction is spontaneous as it increases the entropy of the universe. Figure 7.2 considers the case in which the opposite is true.

Figure 7.2: Enzymes Couple Reactions

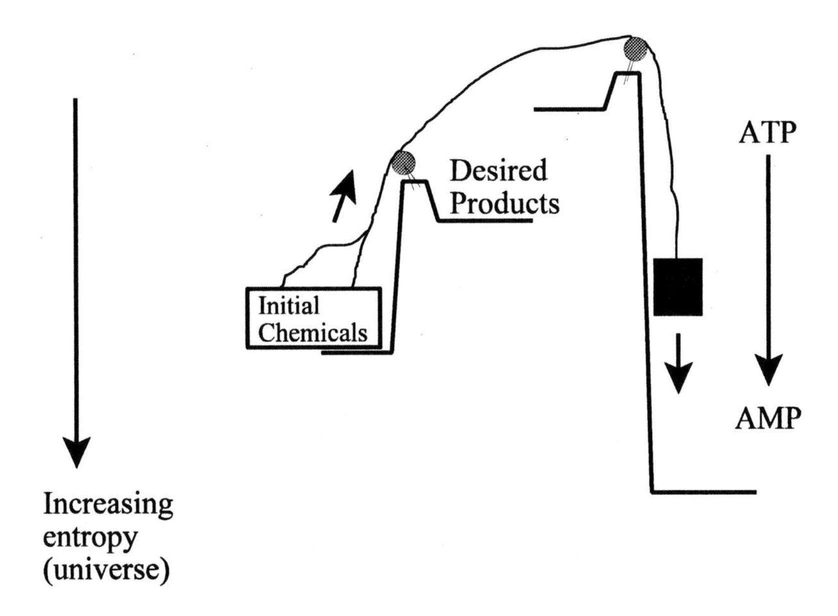

In figure 7.2, a chemical reaction that increases the entropy of the universe is coupled to one that decreases it. The favorable chemical reaction turns the chemical, ATP into AMP. This reaction is represented by the falling weight. It is attached by a rope to the unfavorable reaction. The rope and pulleys represent the enzyme.

Life requires both the techniques shown in 7.1 and 7.2 to maintain its state so far from equilibrium, and the technique shown in 7.2 requires a continual input of energy. This is why life requires food. Sugar is required by animals to create chemicals like ATP. ATP is a very high energy chemical, and when it reacts with water to form AMP, the entropy of the universe is greatly increased. This reaction is coupled by enzymes to many unfavorable chemical reactions.

The two previous examples show how enzymes implement procedural knowledge. The knowledge may also be conditional (figure 7.3).

Figure 7.3: Enzymes Can Make Decisions

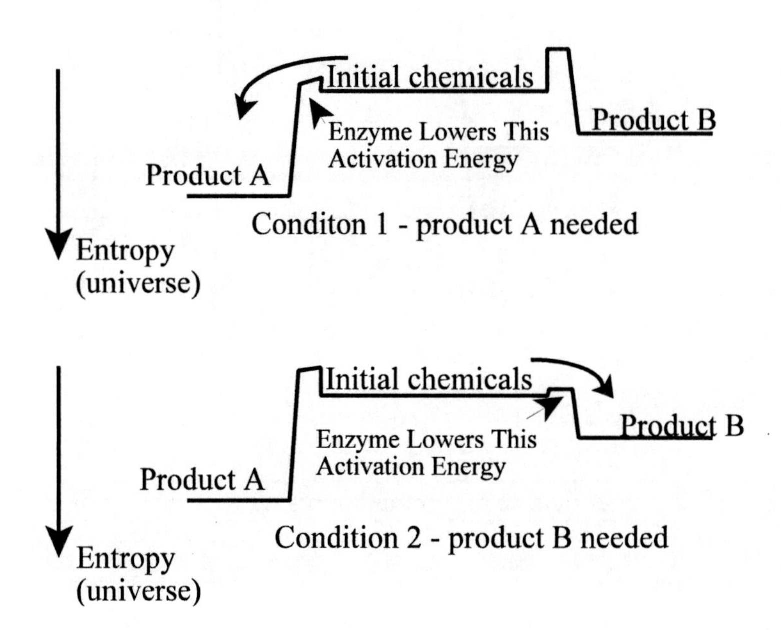

Conditional knowledge is very powerful. It allows life to change when the environment changes. Life does not have to think about how to do this. The decision making is preprogramed just like a computer. Conditional knowledge is not limited to enzymes. Often proteins interact with sections of DNA to promote or repress gene expression. The effects of conditional knowledge can be seen in any higher form of life. Conditional knowledge determines whether a cell becomes a skin, liver, muscle, lung, kidney, or stomach cell.

The last trick used by life is the most subtle, but yet at the same time the most powerful. Life uses teams of enzymes that work together to create desired products, and in many cases, the enzymes that do this do not even need an energy source to accomplish their goal, see figure 7.4.

Figure 7.4: Team of Enzymes Working Together

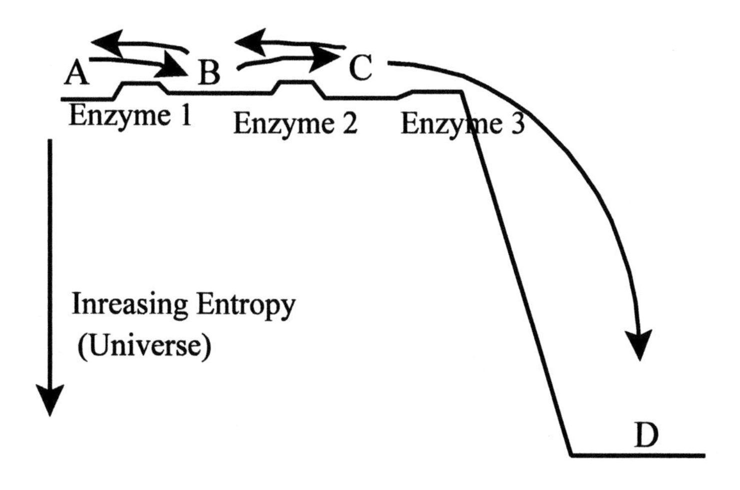

In figure 7.4, enzyme 1 lowers the barrier enabling A to form B, but because the change in entropy is small, the reaction proceeds in both directions. The same is true for enzyme 2, and product C, but these first two reactions are coupled to a third, in which C changes into D. This reaction only proceeds in the forward direction because of the large increase in entropy; as a result, the chemicals, A, B and C, will all eventually be converted to D, and only D will exist. Figures 7.1 to 7.4 illustrate how proteins (enzymes) implement the knowledge contained in DNA.

To avoid violating the second law, life requires quite a bit of complexity. Hundreds of enzymes work together with DNA to decide which chemical reactions take place and then make sure that only the desired chemical reactions happen. Enzymes alter equilibrium by speeding up desirable reaction pathways compared to undesirable ones, and if they need to, they will couple unfavorable reactions to favorable ones forcing the unfavorable reactions to happen and happen quickly, and all of this is subject to the built in ability to make decisions as to which chemical reactions are appropriate at any given time.

Hundreds of enzymes make it possible for life to exist in a state of very low entropy. Most origin of life theories concentrate on a single self replicating chemical. What is not clear is how such a chemical can self replicate without violating the second law. Such a chemical does not have a hundred enzymes working in concert on its behalf to help it self replicate. Before the second law was understood, many scientists tried to build machines that would run forever (perpetual motion machines). They all failed because the second law does not allow such machines to exist. A self replicating molecule is very similar to a perpetual motion machine. If one day, a self replicating molecule is created in a lab, it will replicate itself for a very short period of time and then cease.

Life is possible because life can control and direct the flow of entropy. In other words, life uses energy sources to perform work. When sunlight hits a black pavement, most of its energy is converted into heat. This process greatly increases the entropy of the universe. When sunlight strikes a plant something else happens. Plants know how to harness the energy in sunlight to do work. While much of the sunlight is still converted to heat, some of it is used to make sugar in a process called photosynthesis. Photosynthesis also increases the entropy of the universe. So the second law does not prohibit the process. The entropy increase in photosynthesis is certainly less than that if the sunlight had struck a black pavement, but both processes are allowed because both create more entropy. Plants possess molecular knowledge, and this knowledge enables them to harness the energy of the sun to do work.

Fire is a chemical reaction in which oxygen combines with complex organic chemicals to create carbon dioxide and water. The heat released by a burning bush increases the entropy of the universe, and it does so very quickly, but something else happens when a deer eats the bush. The leaves are eventually converted into carbon dioxide and water, but the energy released by this process is harnessed to perform work. It is used to create ATP. ATP is then used to drive many unfavorable chemical reactions (figure 7.2). The deer does not have to think about how to do this. The knowledge is built into the molecules that make up the deer.

Is Life Really Different from Non-life?

Proponents of a naturalistic explanation routinely appeal to the fact that the earth is an open system. Thus, the sun provides the energy for systems to maintain their distance from equilibrium (existing in a low entropy state indefinitely). While this argument is certainly true for life, it is not true for the chemicals residing in a puddle somewhere on earth 4 billion years ago. *Without a team of enzymes working together with the common goal of self replication, energy sources do not help. There is simply no way to harness the energy to achieve the goal.*

Consider a man and his car with no gas in the desert. The man is not concerned. He knows that he is in an open system and that he has an unlimited supply of energy in sunlight. He theorizes that after the car absorbs enough sunlight, it will suddenly start and he can drive home. His theory is flawed because his car does not know how to use the energy from the sun to perform useful work, and 5 billion years will not solve his problem.

Do Energy Sources Really Help?

Both experimental and theoretical evidence suggests that sunlight and other energy sources make it harder for chemicals to exist far from equilibrium. This is easily understood in terms of activation energies. A low entropy state may exist indefinitely as long as the chemicals involved never obtain enough energy to cross the activation barrier, but such states are very hard to maintain when energy is plentiful. Energy sources make the origin of life even more difficult to explain. Why so many authors continue to point out their benefits is hard to understand.

In order for a self replicating molecule to replicate indefinitely, it must possess a mechanism to harness and use energy. Self replication decreases the entropy of the universe, and as such, it always requires a source of energy. Therefore, unless a self replicator possesses both the knowledge and ability to drive its replication with a plentiful energy source, the molecule's very existence violates the second law.

Non-Equilibrium Thermodynamics

The idea that energy sources can somehow solve the origins problem is so prevalent that a few detailed examples are required. A flow of energy can maintain a living or a non-living system in a state of non-equilibrium. Consider a chemical reaction that proceeds as follows: if the temperature is less than 25 degrees Celsius, then C –> A + B. If the temperature is above 100 degrees Celsius, then the reaction proceeds in the opposite direction, A+B–> C. Now consider a body of water that is warmed by volcanic activity (figure 7.5).

Figure 7.5: A Non-Equilibrium System

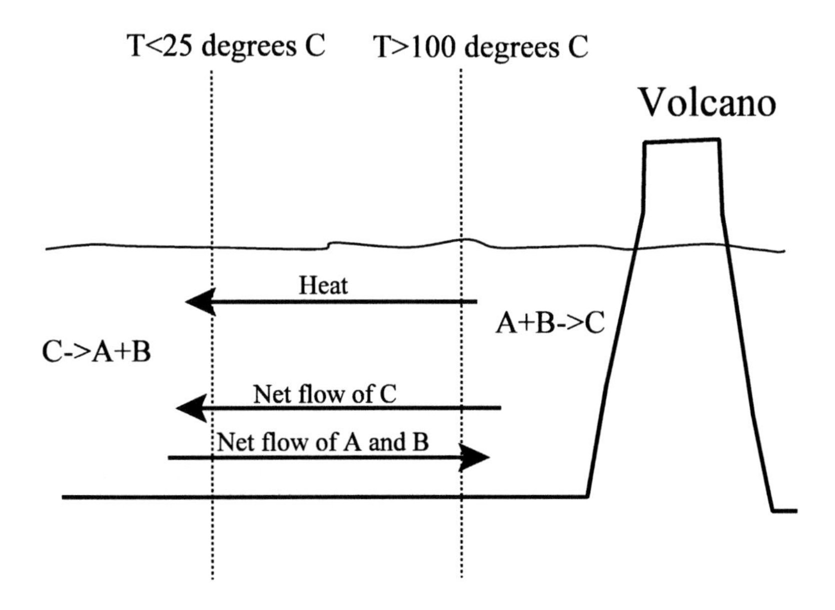

In figure 7.5, there is a flow of heat from the volcano into the ocean around it. This sets up a temperature gradient. The water is very hot near the volcano and very cool away from it. Heat flows in the water as indicated by the arrow. The chemical reaction under consideration proceeds in the direction to create chemical C near the volcano. Away from the volcano, it proceeds to create the chemicals, A and B.

Since this creates much more of chemical C near the volcano, C tends to move away toward the colder water (see arrows). Much more of the chemicals A and B exist in cold water. So these tend to travel toward the warmer water near the volcano (see arrows). This creates a cycle in which some chemicals are continually transported from warm water to cold water, and others from cold to warm.

If the desired chemical is C, then it would be beneficial to heat the entire body of water to a temperature greater than 100 degrees. If the desired chemical is A or B, then it would be beneficial to have the entire ocean at a temperature less than 25 degrees Celcius. The flow of energy allows the chemicals to exist away from equilibrium, but they exist in an intermediate state between the two extremes. There is less of the chemical C than if the entire body of water is heated to 100 degrees Celsius, but there is more C than if the entire ocean is at 25 degrees Celsius. The same argument applies to A and B.

The relevance of such a system to the origin of life is questionable at best. How this flow of energy and matter can create a complex biological molecule is not clear. Figure 7.4 is certainly a form of order. C flows to the left, and A and B flow to the right. The cycle is maintained by the flow of heat, but there is nothing if figure 7.4 that performs the function of an enzyme. There is no mechanism to couple an unfavorable reaction to a favorable one. Furthermore, there is no mechanism to preferentially create only the desired chemicals. But most importantly, there is no way to store information. So there is not way for such a system to evolve.

Experimental evidence supports this theoretical conclusion because most if not all experiments designed to investigate the origin of complex chemicals make extensive use of non-equilibrium conditions. The most famous, Miller's electric discharge experiment, will be discussed in chapter 9.

So in this case, both theory and experiment converge to the same answer. Under plausible prebiotic conditions, without molecular knowledge, it is very difficult (if not impossible) to create the complex chemicals used by life today.

Chemical Oscillators

Chemical oscillators are systems of chemicals that exhibit very interesting time-based fluctuations when they are far from equilibrium. These fluctuations may cause a solution of chemicals to change colors in a periodic fashion or create complex spatial patterns (figure 7.6).

It would be incorrect to suggest that these systems possess molecular knowledge because their behavior does not benefit any of the chemicals involved. They do not self replicate. Nevertheless, they are interesting anomalies. On their way toward chemical equilibrium, these systems take a detour and veer off course. They eventually reach equilibrium, and the oscillation stops.

These systems do not help explain the origin of life. They have never been implicated in the prebiotic synthesis of proteins, DNA or RNA. *The chemicals in these oscillators cannot evolve because they have no mechanism to store information, and they do not self replicate.* These systems are like a man-made machine. For example, a battery powered watch is a chemical oscillator. The watch hands move in a periodic fashion, and the hands stop when the battery dies. The knowledge that enables the watch to do this is built into it by engineers. In a similar fashion, chemists bring together the necessary chemicals for chemical oscillators.

Figure 7.6: A Chemical Oscillator

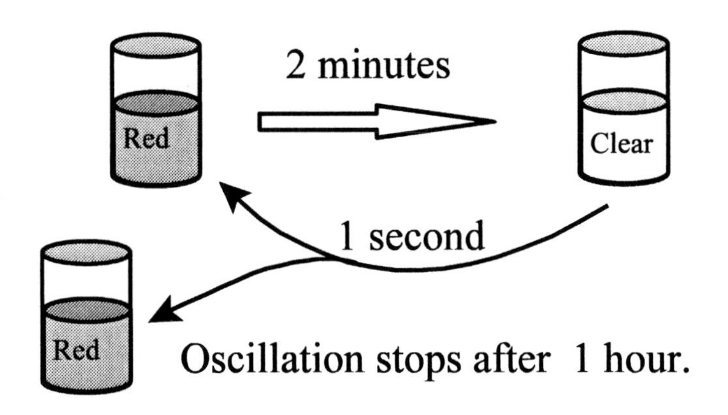

The popularity of chemical oscillators in the origins field rests largely on the shoulders of two authors, Kauffman and Prigogine. Both dedicated an entire book to this one subject. Despite their enthusiasm, chemical oscillators do not solve any of the problems related to the origin of life. While they do tap an energy source to perform work, the energy source is limited, and once it runs out, the oscillation ceases. Furthermore, because these systems have no way to store molecular knowledge, they cannot evolve, and perhaps most importantly, the systems do not self replicate. A chemical oscillator has about as much chance of evolving into a living organism as does a watch. Neither possesses the knowledge or has the ability to acquire the knowledge required for self replication. So natural selection cannot optimize these systems, and evolution cannot take place.

Kauffman has also put forth several other ideas concerning the origin of life and self organization. For example, in order to create a self organizing system, he envisions a set of chemicals that perform the many functions of enzymes described in this chapter. So he defines a self organizing system as life, and it stands to reason that life can self replicate. He just assumes that the molecular knowledge and a method to implement the knowledge already exist, and once these two are in place, life is inevitable. Unfortunately, his ideas do not explain how the knowledge arose in the first place. He also fails to consider the need to tap a plentiful energy source to perform work. So most of his ideas give rise to perpetual motion machines.

Furthermore, he continually confuses order with knowledge. Because most of his thought experiments and computer simulations actually give rise to order, many of the ideas that he proposes are not relevant to the origin of life. For example, he simulates the order that arises when many logic gates are connected together in a random fashion. He observes large sections of these arrays oscillating at the same frequency. Any electrical engineer would immediately recognize his system as nothing more than a collection of ring oscillators, and the order that results contains no useful information.

Entropy and Biological Evolution

Several authors have suggested that the second law works against evolution. While this is certainly true for chemical evolution, the second law does not prevent existing biological systems from evolving. If anything the contrary is true. Evolution is possible because life does not seem to be able to copy its DNA without making an occasional mistake. If DNA was replicated faithfully in every generation, then chance would never create new useful information, and natural selection would never have the opportunity to preserve anything new.

The second law is often stated as the disorder of the universe must always increase. This statement is not true, because entropy is a measure of uncertainty not disorder. Several authors have used the entropy as disorder definition, to justify the conflict between the second law and evolution. Since life is not disordered, it is possible to see the conflict when entropy is defined in this way. Nevertheless, entropy and disorder are not the same.

The second law stipulates that the uncertainty of the universe will increase with time. This uncertainty is reflected in all physical systems. Mutations increase uncertainty. Thus, changes to existing genes and proteins are fully expected and consistent with the second law.

The direction toward increasing complexity that life displays with time may be attributed to the preserving power of natural selection. If chance creates useful information that confers a selective advantage, then natural selection will preserve it. This process has nothing to do with the second law, and authors who suggest that the second law and evolution are somehow mutually exclusive do not understand the nature of entropy.

References:

1) Morowitz, Energy Flow in Biology, Ox Bow Press, 1979.
2)Brillouin, Science and Information Theory, 1956.
3) Prigogine, Stengers, Order Out of Chaos, Bantam Books, 1984.
4) Kauffman, At Home in The Universe, Oxford University Press, 1995.

Chapter 8: The Structure of DNA, RNA and Proteins

This chapter will explore the structure of nucleic acids and proteins.

DNA Structure

DNA is composed of several different subunits. The backbone of the molecule is made of a sugar called deoxyribose. The deoxyribose is held together by phosphate groups. Deoxyribose also forms bonds with the four bases, adenine (A), cytosine (C), thymine (T) and guanine (G). Figures 8.1-8.6 depict how the subunits are assembled in a DNA molecule. In these figures, black represents adenine, white thymine, dark gray guanine, and light gray cytosine.

Figure 8.1: Rasmol Image of DNA Double Helix Segment

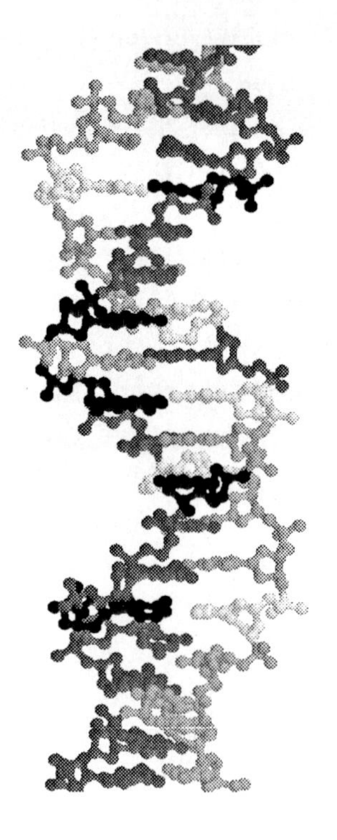

Figure 8.2: Closer View of DNA

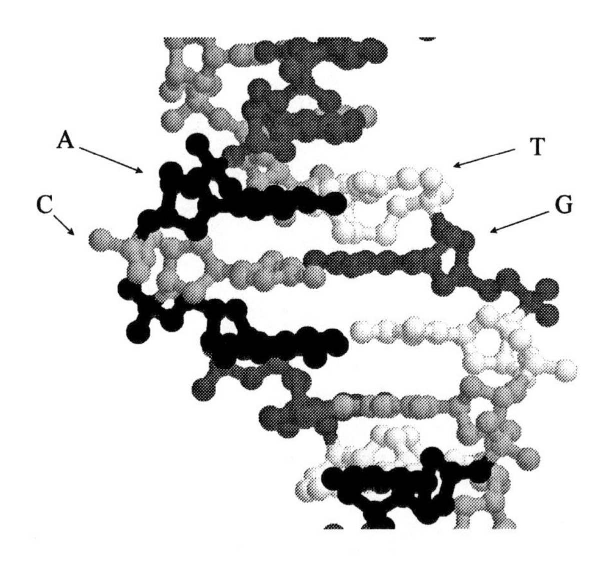

Figure 8.3: Conceptual DNA Model

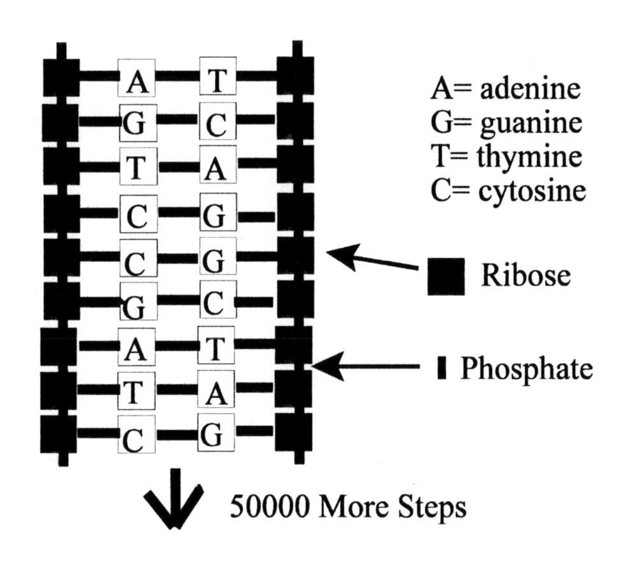

Figure 8.4: Top and Side view of Two Steps

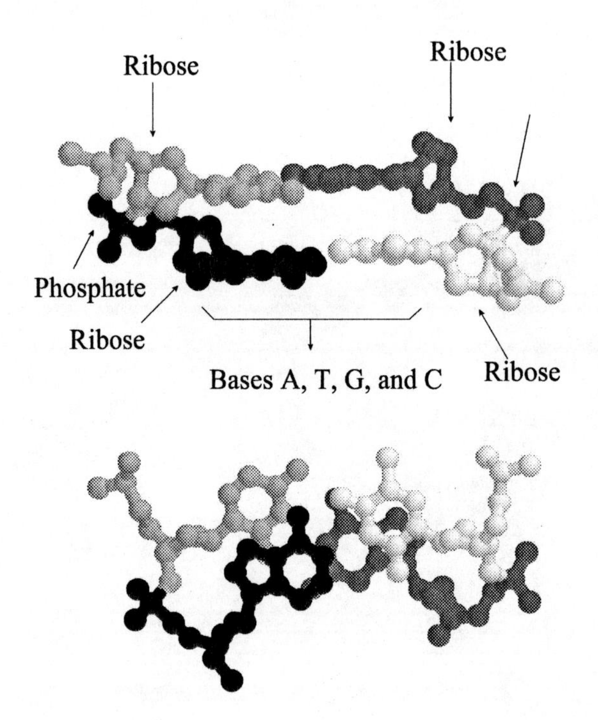

Chapter 3 describes how this structure stores information. The order of the four bases (A, T, C and G) read three at a time per table 3.2 determines the amino acid sequence in the final protein.

RNA Structure

RNA is very similar to DNA, but it normally does not form the characteristic double helix (twisted ladder). RNA is a mixture of single stranded and double stranded segments (figure 8.7). The 3-D structure is often stabilized by complementary base pairing in short regions (boxes in figure 8.7). RNA uses the base uracil in place of thymine. It also uses ribose instead of deoxyribose (figure 8.6).

Figure 8.5: Detailed DNA Structure

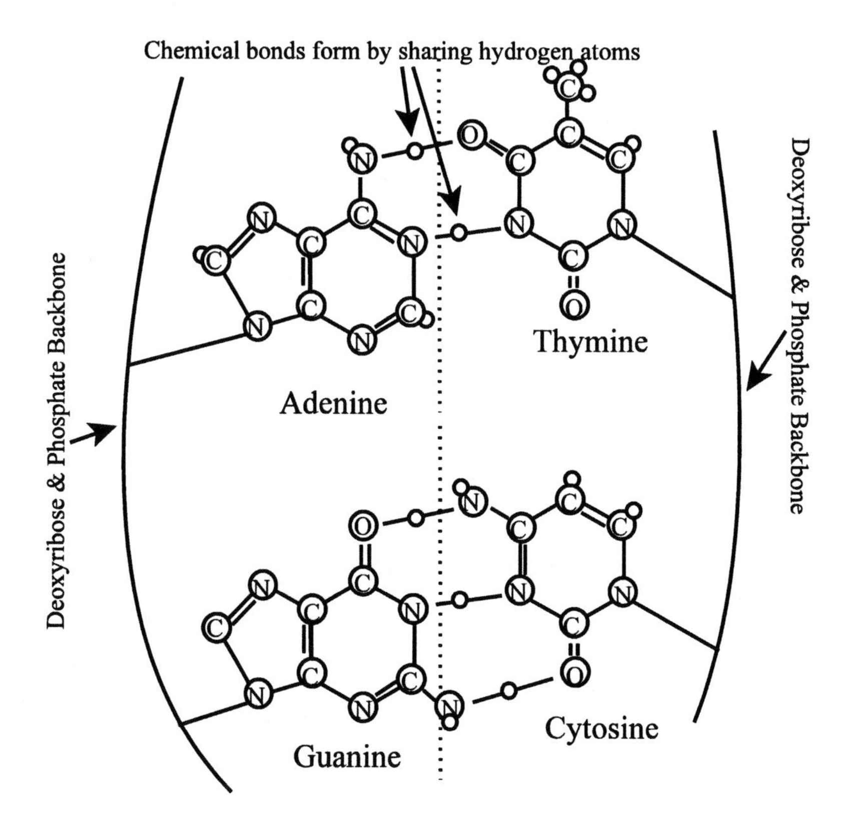

Figure 8.6: Ribose, Deoxyribose, Uracil and Thymine

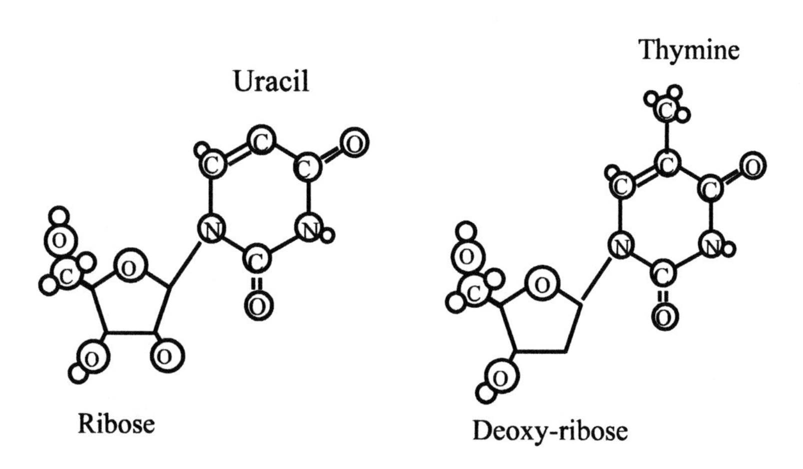

The 3-D structure is extremely important, because it conveys to RNA some of the properties found in proteins. RNA can regulate chemical reactions just like proteins. This is why RNA has taken center stage in most origin of life theories. RNA can store information and it can regulate chemical reactions. Nevertheless, the initial optimism for self replicating RNA molecules has largely been replaced by doubt. The reasons for the doubt will be discussed in the next two chapters.

Figure 8.7: RNA Structure Stabilized by Base Pair Bonds

RNA Curls up On Itself Form Double and Single Stranded Regions

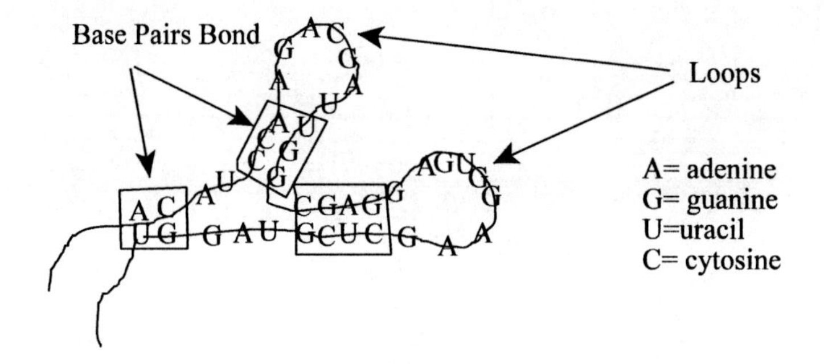

Proteins

This section explores the structure of proteins and amino acids.

Organic Chemistry Functional Groups

Organic chemistry is the study of how large chemicals containing carbon interact. Because organic chemicals are very large, the structures of many are overwhelming. Fortunately, the chemicals can only interact with each other in a very limited set of reactions. These reactions are controlled by functional groups. The functional groups relevant to the origin of life are shown in figure 8.8. Carboxylic acid and amino groups are found in all amino acids. P = phosphorous, O = oxygen, N = nitrogen, C = carbon, and unlabeled small spheres = hydrogen.

Figure 8.8: Functional Groups

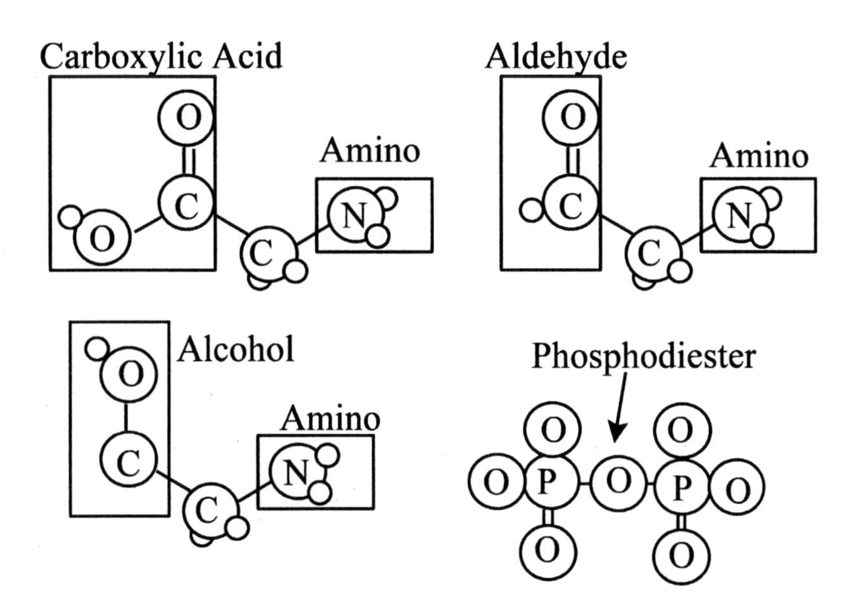

Structure of Amino Acids

Amino acids are composed of 5 elements - carbon, oxygen, nitrogen, hydrogen and sulfur. It is their chemical structure that permits them to form long chains. These long chains fold up into complex three dimensional patterns forming proteins. Each amino acid can be broken down into three critical parts: the side chain or side group, the N-terminus (amino group) and the C-terminus (carboxylic acid). The N-terminus of one amino acid can attack the C-terminus of another. Under the right conditions, this attack will form a link between the amino acids. This link is called an amide bond or a peptide bond. Figure 8.9 illustrates the key structural components using the amino acid glycine.

Figure 8.9: Glycine and the Peptide Bond

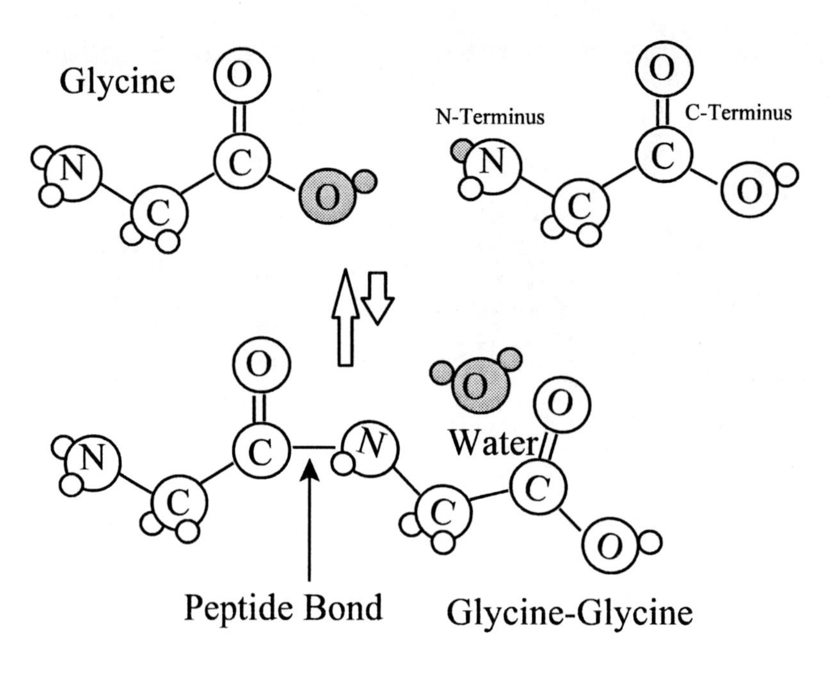

Notice that after the peptide bond forms, and the two glycines are linked by a peptide bond, one C-terminus and one N-terminus still exist. This allows the chain to continue growing. A short chain of amino acids is called a peptide. A peptide may or may not contain information. The dark atoms in figure 8.9 are the atoms that leave forming water when the two glycine molecules join. Because this reaction creates water, it is called a condensation reaction. Condensation reactions do not occur readily in water and are particularly problematic for RNA and DNA prebiotic synthesis.

Figure 8.10: RNA and DNA Requires Many Condensation Reactions

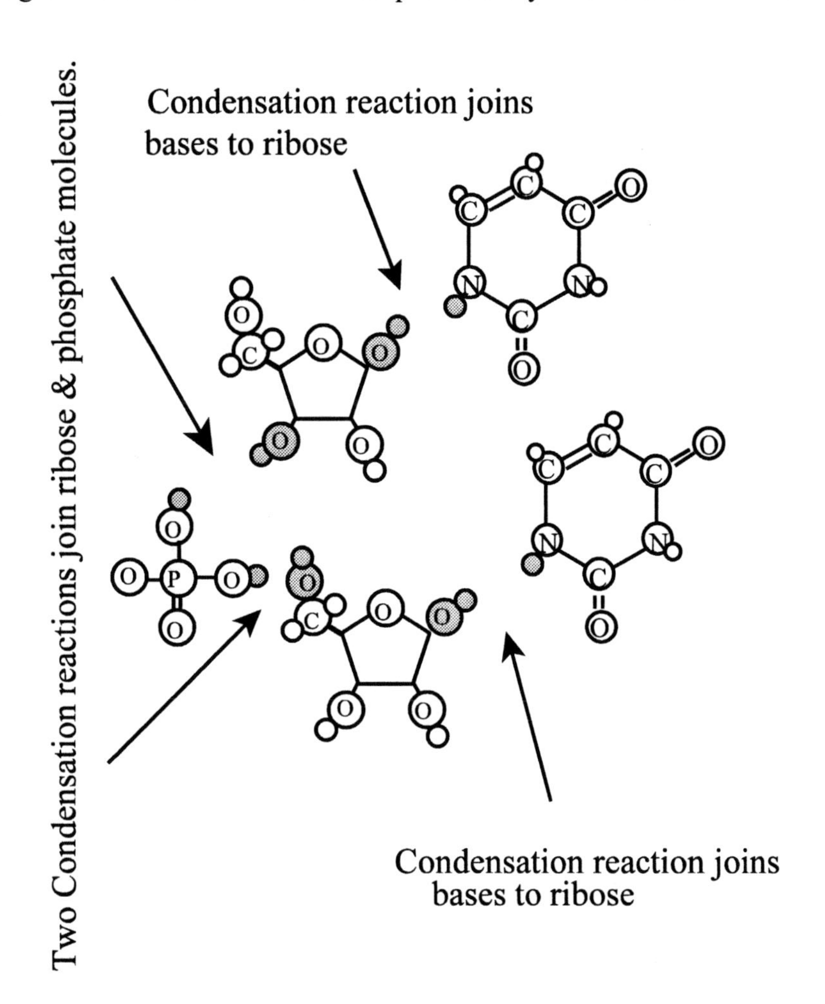

A protein is a chain of many amino acids (typically more than 100 amino acids). Proteins have a specific function. Therefore, proteins contain useful information (knowledge). This information is specified by the order of amino acids in the chain. The 20 amino acids used by life differ only in their side chains.

Every amino acid except glycine has two forms, L and D. One form is the mirror image of the other. These forms are called isomers. The L and D isomers of the amino acid alanine are shown in figure 8.11.

Figure 8.11: L and D Isomers of Alanine

These two molecules are mirror images.

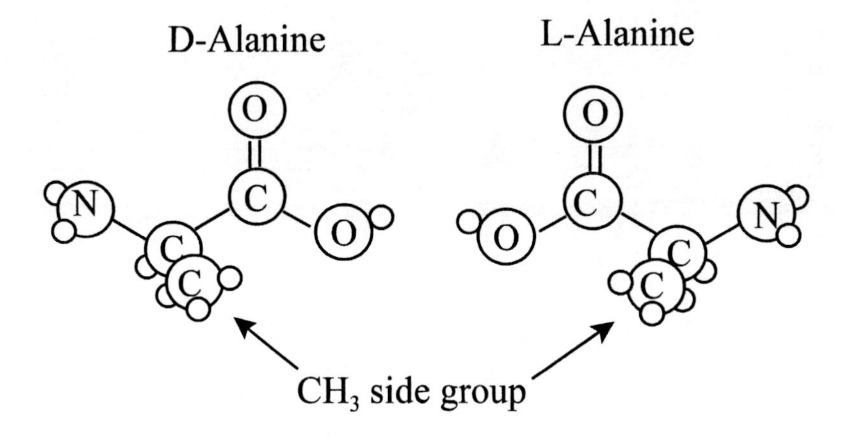

L-Alanine is used by life. D-alanine is not.

The L and D isomers can form peptide bonds. But the location of the side group is located in the wrong position when the L isomer is replaced with the D. This of course may influence protein function. One of life's greatest mysteries is why did life chose to only use the L-amino acids? Nineteen of the amino acids used by life are shown in figures 8.12-8.17. Glycine is shown in figure 8.9.

Figure 8.12: Hydrophobic Amino Acids (Do Not Like Water)

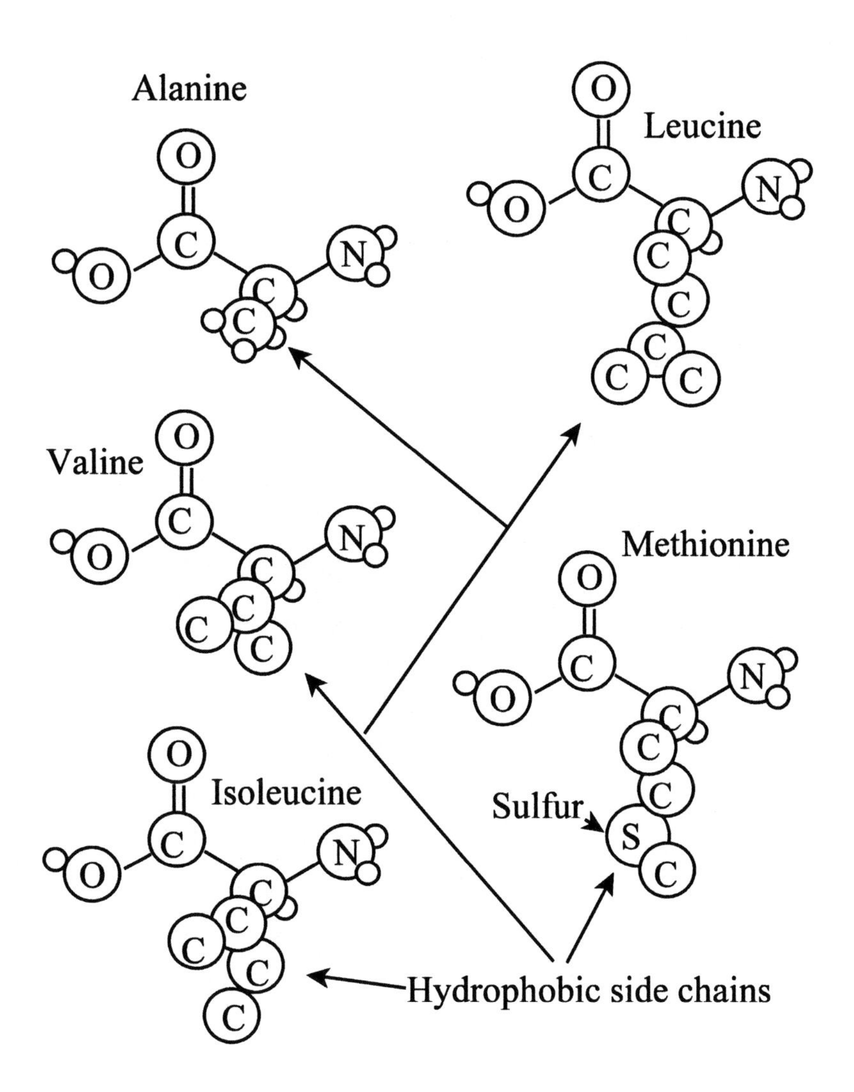

Figure 8.13: Basic Amino Acids

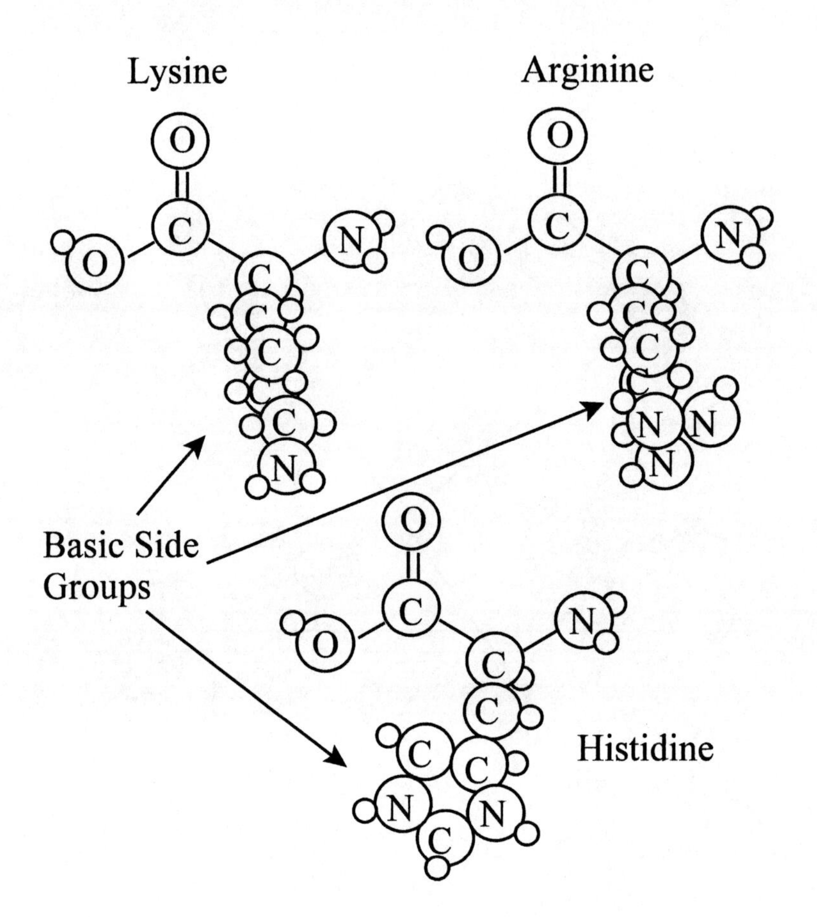

Figure 8.14: Polar Uncharged Amino Acids

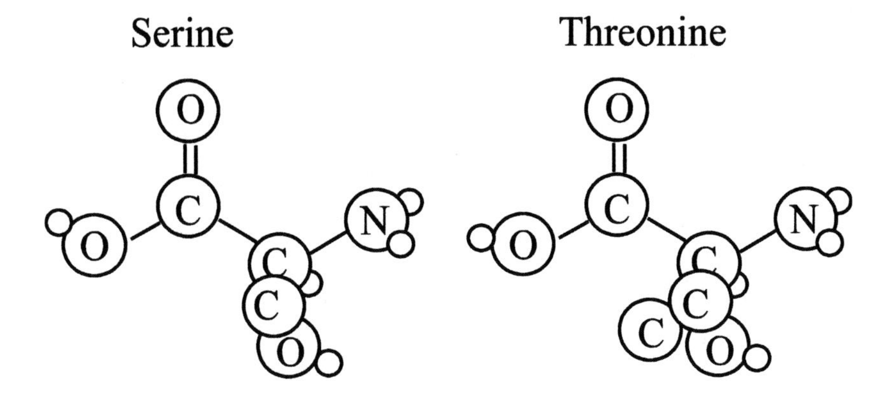

Figure 8.15: Two Unique Amino Acids

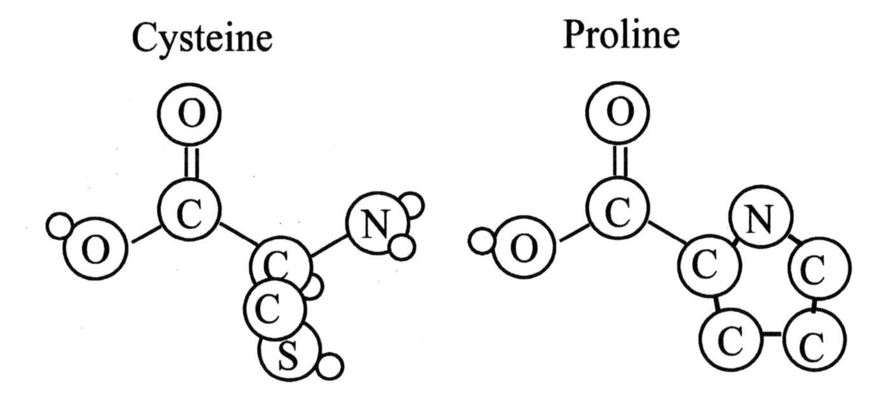

Figure 8.16: Bulky Amino Acids

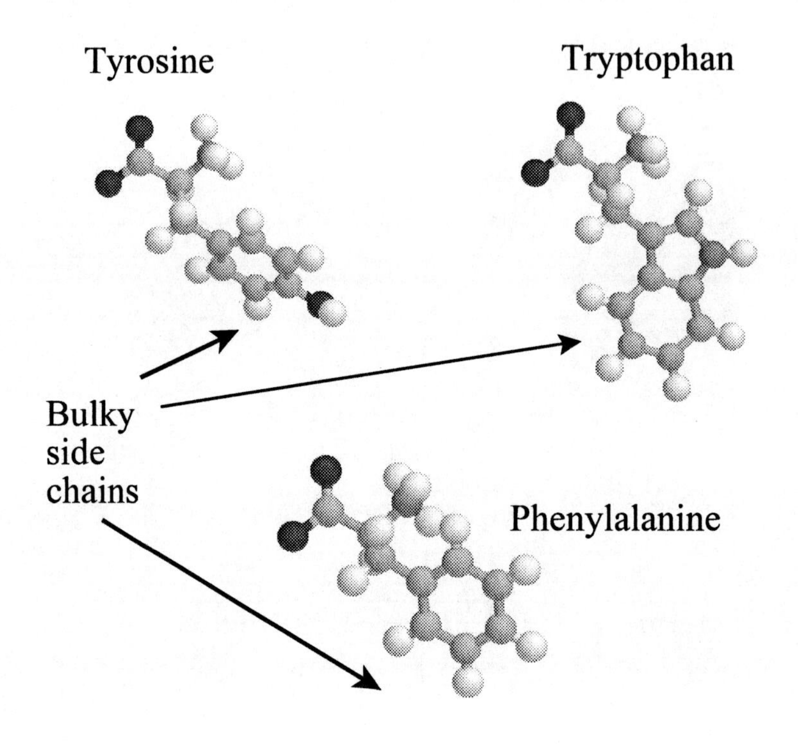

Figure 8.16 was generated using Rasmol. This is a free molecular viewer that allows users to view the 3-D structure of complex molecules. In this view, oxygen is black, hydrogen is white, carbon is gray, and nitrogen is a slightly darker gray. Rasmol was also used to generate the DNA images at the beginning of this chapter.

Figure 8.17: More Rasmol Images of Amino Acids

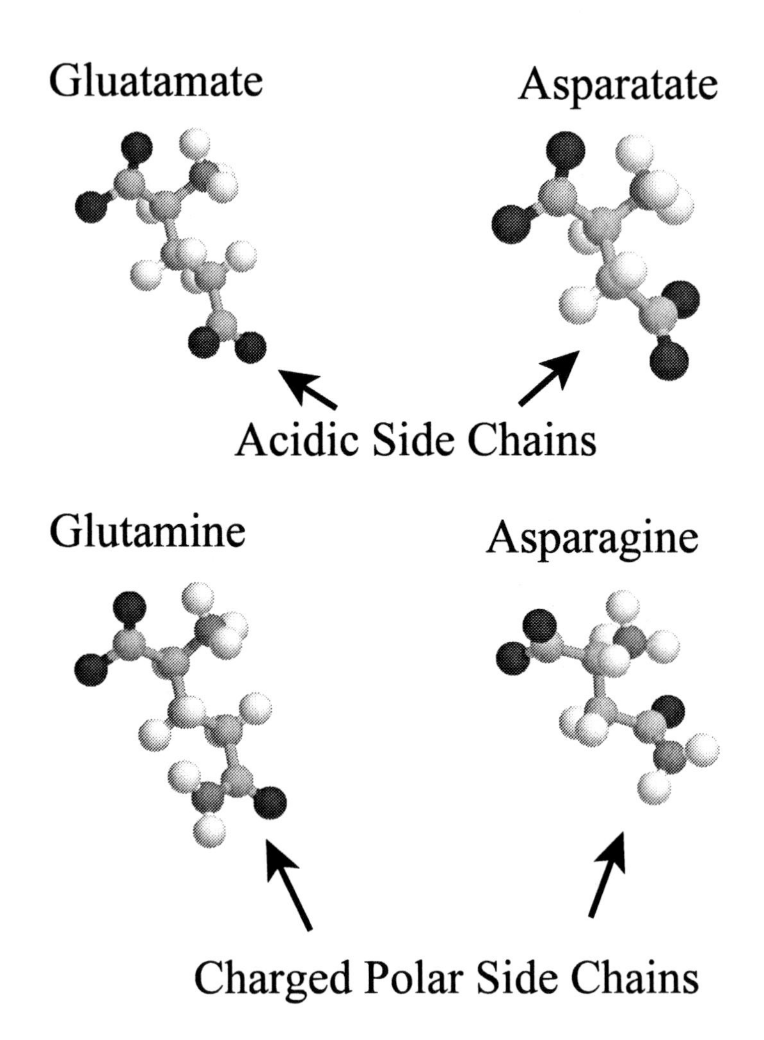

Chapter 9: Prebiotic Synthesis of RNA, DNA and Peptides

Naturalistic theories concerning life's origin began to take shape in 1953. Watson and Crick unraveled the structure of DNA, and Stanley Miller performed an experiment showing that amino acids can be produced in a spark chamber. Most scientists of the day assumed that the mystery of life's origin would be solved in a few years.

The early pioneers in this field realized that a complete living organism, like the bacteria in figure 9.1, could not spontaneously appear in a spark chamber or in any other environment governed by purely naturalistic laws. The pioneers needed the first form of life to be simpler than any living thing that is present on earth today.

Figure 9.1: Information Transfer in a Bacteria Cell

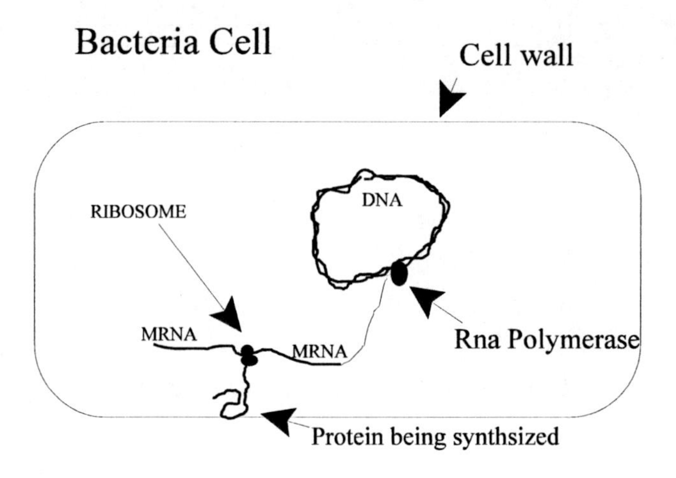

Initial theories hypothesized that the first living thing was a protein. These theories seemed reasonable at the time because many of the building blocks of proteins, amino acids, are easily synthesized under plausible prebiotic conditions. Because proteins regulate and control almost all of the activities necessary for life, the living protein theory quickly gained widespread acceptance, but soon scientists realized that there was a major flaw with the protein theory.

Proteins cannot self replicate, so the first living protein would not be able to reproduce itself, and without replication there can be no natural selection; therefore, the first living protein would have no way to evolve.

This issue led to the demise of the protein theory. In its place, emerged the RNA theory. This theory gained substantial momentum when it was found that just like proteins some RNA molecules can catalyze chemical reactions. Recently this theory has also fallen out of favor because it has its own set of problems which will be discussed later. Today the most popular theory involves a self replicating pre-RNA molecule.

Self replicating molecules are probably not the best theory to pursue, because such molecules cannot reproduce for any length of time without running into serious problems with the second law. Nevertheless, many researchers in the origins field are absolutely sure that the first living thing was a self replicating chemical, and their point of view is understandable. There is simply no chance that a complete bacterium spontaneously formed from the chemicals in a puddle four billion years ago. In many ways, a self replicating molecule that violates the second law is a better choice.

Nevertheless, the second law should not be casually dismissed because its existence explains why investigators have not been able to create a self replicating molecule in the lab. Unless a self replicator has the knowledge and ability to harness the power of sunlight (or some other abundant energy source) and use this energy to drive its own replication, then its lifetime will be short lived and its existence forbidden by the laws of physics.

The origin of self replication requires a solution to five problems:

- Chemical evolution must create a protein, an RNA molecule or an RNA like molecule.
- This molecule must possess the molecular knowledge that enables self replication.
- It must also be able to implement this knowledge.
- The molecule must be able to harness an energy source to do useful work.
- The first self replicator must be able to synthesize any chemicals lacking from its surroundings that it needs to self replicate.

Experiments investigating the origin of life have for the most part ignored the last two issues. This is understandable because until a molecule that can at least replicate itself for a little while can be found, there is no need to try to find one that can replicate itself indefinitely. This chapter will investigate the prebiotic synthesis of RNA and proteins. The next chapter will investigate self replication.The pioneers in chemical evolution expected to show that the primordial ocean was full of biological molecules. These researchers suggested that the early atmosphere contained no free oxygen, and that under these conditions, the required biological precursors should be plentiful. The remainder of this chapter will evaluate the validity of this hypothesis.

It is extremely difficult to synthesize biological molecules under plausible prebiotic conditions, and today this difficulty has led most to conclude that the primitive ocean contained a very limited supply of biological precursors. This finding does not mean that the primordial soup did not exist. It does mean that the primitive ocean was not the primordial soup because any relevant molecules in it would be too dilute.[4,11,18]

It is possible to imagine environments that will concentrate biological precursors, but this leads to further problems. It limits the soup in such a way that the conditions necessary for its existence rarely exist and leads to the perhaps alarming conclusion that even given 5 billion years the soup may not have existed.

Zero Tries

The goal of this chapter and the next is to show that given 5 billion years and an almost unlimited source of energy, the probability of creating a protein or an RNA molecule is vanishingly small. Furthermore, the probability that the molecule so created contains the knowledge needed to self replicate is also vanishingly small. The chance of success is given by multiplying these two vanishingly small numbers. The trapped scientist in figures 9.2 and 9.3 helps illustrate this concept. With zero tries, even a short combination eludes the scientist (figure 9.2), and unfortunately, the required combination for self replication whether protein or RNA is quite long (figure 9.3).

Figure 9.2: Zero Tries

Figure 9.3: Zero Tries with Long Combination

Investigator Interference

In figure 9.2, the scientist is not cooperating. He refuses to play the game. He enters no words into the computer, so he accumulates no tries. The researches are very unhappy with these results. So they blast the door with dynamite. This of course opens the door (figure 9.4). The researchers then conclude that given 5 billion years the door will open. Figure 9.4 is an obvious example of investigator interference.

Figure 9.4: Investigator Interference

The concept of investigator interference was first introduced by Thaxton et al. in The Mystery of life's Origin: Reassessing Current Theories. In this book, the authors suggest that some interference is warranted. Scientists cannot conduct experiments that last for one billion years. So interference is useful in that it speeds up the process of evolution, and to be fair, the interference is a great learning tool because it allows scientists to rule out extremely unlikely scenarios. Thaxton also concludes that in many cases the interference is excessive.

While the interference is a good idea because it helps scientists learn, it can also be very misleading. The scientist did not open the door in figure 9.4. The dynamite opened the door. Any conclusion that given time, the scientist will open the door is completely unfounded. This chapter will introduce many examples of interference. Readers should use their own judgement as to whether the degree of interference is acceptable or excessive using the following criteria: if the artificial conditions generated in the lab might happen in nature given 5 billion years, then the interference is acceptable. Otherwise, it is excessive. Proteins will be considered first followed by RNA.

Protein Synthesis

Synthesizing proteins under prebiotic conditions is not as straight forward as many would have predicted. Ten (maybe 12) of the amino acids are relatively easy to create. Both L and D isomers are created, and two amino acids alanine and glycine almost always dominate the mixture. Despite these issues, creating amino acids is not that difficult. It is forcing the amino acids to form peptide bonds that is difficult.

Miller's Experiment

Figure 9.5 illustrates the Miller spark chamber. The water in the flask is boiling. The atmosphere above the water and in the spark chamber is controlled. In this example, hydrogen, methane and ammonia are introduced. The electrode is charged to a very high voltage, and it creates an electric spark. This spark is an energy source. It allows the chemicals in the chamber to react and form new chemicals. The condenser removes the chemicals from the spark chamber, and they accumulate in the trap. Life uses 20 amino acids. Miller's chamber can create between 0 and 10 of the 20 (the number created depends on the gases used in the atmosphere).

The chamber also creates many other chemicals. Other scientists have repeated this experiment with alternative energy sources like UV light and heat. These experiments demonstrate that many amino acids are easy to create. Miller's chamber is a non-equilibrium system cleverly designed and optimized to create nonvolatile organic compounds like amino acids.

Whether or not this experiment is representative of the conditions on the early earth is questionable. Many scientists today do not believe that ammonia, hydrogen and methane were present in the earth's early atmosphere, and without at least one of these, no amino acids are produced by the spark chamber.

Figure 9.5: Miller's Spark Chamber

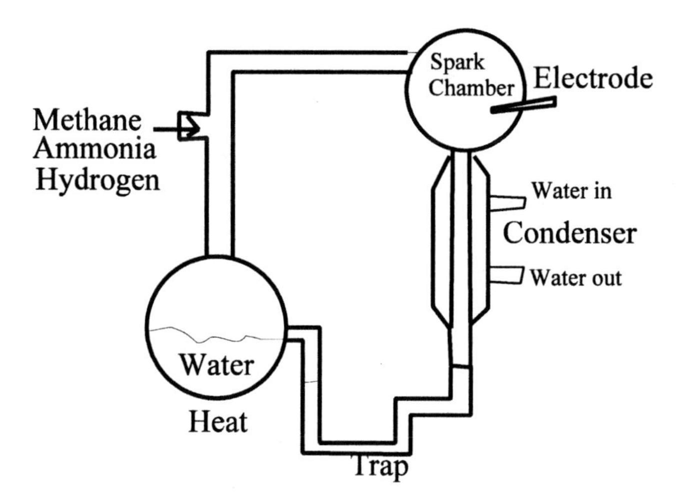

Thermal Proteins or Protenoids

Since water inhibits the formation of peptide bonds, the first step to create a peptide often involves removing water. Fox successfully created chains of amino acids by heating a purified concentration of amino acids to 150 degrees Celsius for about 14 hours. At this temperature, water and other volatile compounds vaporize. This is important because when a peptide bond forms, a single water molecule is also produced. The heat drives this molecule off forcing the reaction forward because without water it cannot go backwards.

Fox obtained very long chains when he included high concentrations of the amino acids, glutamate, aspartic acid and lysine. Fox called the amino acid chains formed by heating, protenoids. They are also called thermal proteins. They are different from normal proteins in two important ways. Thermal proteins contain both D and L isomers, and the peptide bonds that form are very unusual. The side chains associated with lysine, glutamate and aspartate form over ½ of the peptide bonds.[1] This second feature has led most origin of life researchers to drop protenoids as a viable candidate for the first living protein. Stanley Miller in particular has criticized thermal proteins as unlikely candidates because the conditions necessary to form them probably rarely exist. The temperature has to fall within a narrow range (150-180 deg C), and if the heating lasts too long (more than a day), then the thermal proteins are destroyed.[2] Furthermore, given that amino acids will not form thermal proteins without a very high concentration of aspartate, glutamate, or lysine leads to another question. How do proteins with reasonable concentrations of these 3 amino acids form in the soup?

Given 5 billion years, a few thermal proteins may have had a chance to form. In this respect, thermal proteins are unique. While they are not biological precursors (due to the unconventional peptide bonds), they do at least have a chance of existing.

Short Peptides Chains in Water

Short peptide chains have been produced in water. Usually a catalyst like clay or some other mineral like pyrite is required. The minerals promote the formation of peptide bonds.

Peptide bond formation can also be induced by sodium chloride (salt) in the presence of copper.[24] These salt-induced peptide bonds are generally limited to very short chains. Nevertheless, the reaction is of interest because salt and copper are very common, the reaction takes place in water, and at least for a few amino acids, there seems to be a slight preference for peptides composed of L-amino acids. These short peptides may then interact with other minerals like clay to form longer chains.[25] Clay can form pockets that may help exclude water. Figure 9.6 shows how a mineral like clay may help a peptide bond form. The C-terminus of each glycine molecule interacts with the clay substrate. The arrow shows how the N-terminus of one glycine attacks the C-terminus of the other. This forms a peptide bond. Peptide chains of up to 10 amino acids have been created in the lab using these techniques.

Salt, copper and clay are very common. These minerals would have been present in or near the primordial soup and facilitated amino acids joining together into short peptides. While the resulting peptides would be too short to be biological precursors (< 10 amino acids), some of these peptides almost certainly existed on earth before life. A few readers might assume that given 2 billion years these processes would naturally create longer peptide chains. This is unlikely because the soup must have contained chemicals other than the amino acids used by life. Specifically, formic acid, amines, formaldehyde, and non-biological amino acids must have been present, and all of these would interfere with proper chain growth. Furthermore, many destructive processes would destroy any growing chains. So given 2 billion years, maybe a few peptides greater than 10 amino acids evolved, but the quantity would have been very limited. Because the starting point for these experiments model evolution in a test tube, one week may already correspond to a billion years of evolution in nature. This technique certainly did not fill up the primitive oceans with peptides.

Figure 9.6: Clay and Peptide Bond Formation

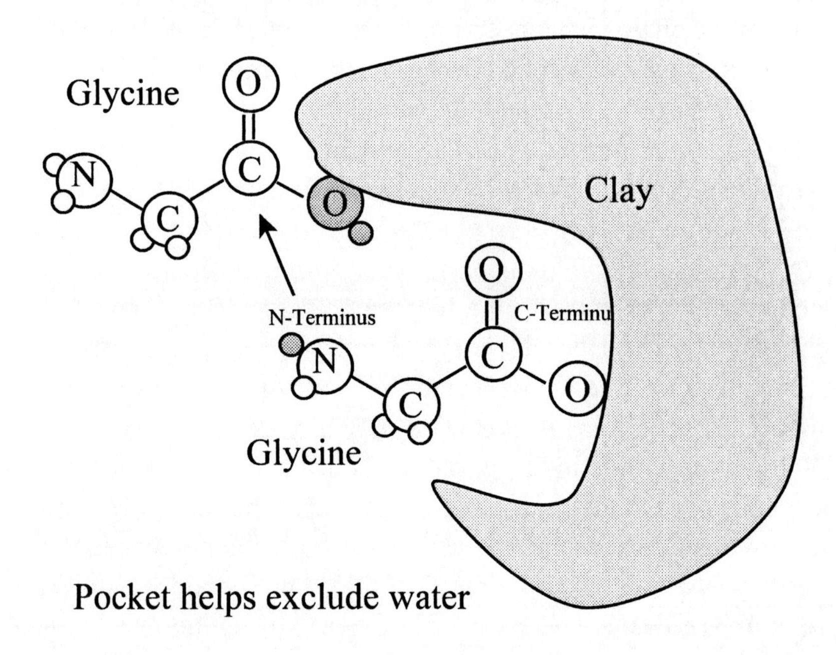

Long Peptide Chains in Water

The last class of experiments will consider extreme investigator interference. In these experiments, a chemical that is either not found in the Miller type spark chamber or is very rare is added to the solution in high concentrations. The chemical is almost always a condensation agent. These chemicals contain double bonds that can absorb water. The reaction is sometimes carried out in a set of sequential steps specifically designed to elongate the peptide chain. Figure 9.7 shows how this process works. The condensation agents must be used in high concentrations because they are not stable in the presence of water. Condensation agents do not have the ability to differentiate between the water molecule released when two amino acids combine, and the water molecules already present in the soup. Thus, finding a condensation agent in the primordial soup is like finding a dry sponge in the ocean.

Many authors have claimed that experiments that use condensation agents are relevant to the origin of life. Nevertheless, the justification for adding a condensation agent that has no plausible route for prebiotic synthesis to the mixture is questionable, and adding it in the excessive quantities required for peptide chain growth is certainly not justified. The sequential washing steps which are controlled by the investigator add to the already excessive interference. The results of these experiments are not relevant to the origin of life.

Figure 9.7: Peptide Bonds With Condensation Agents

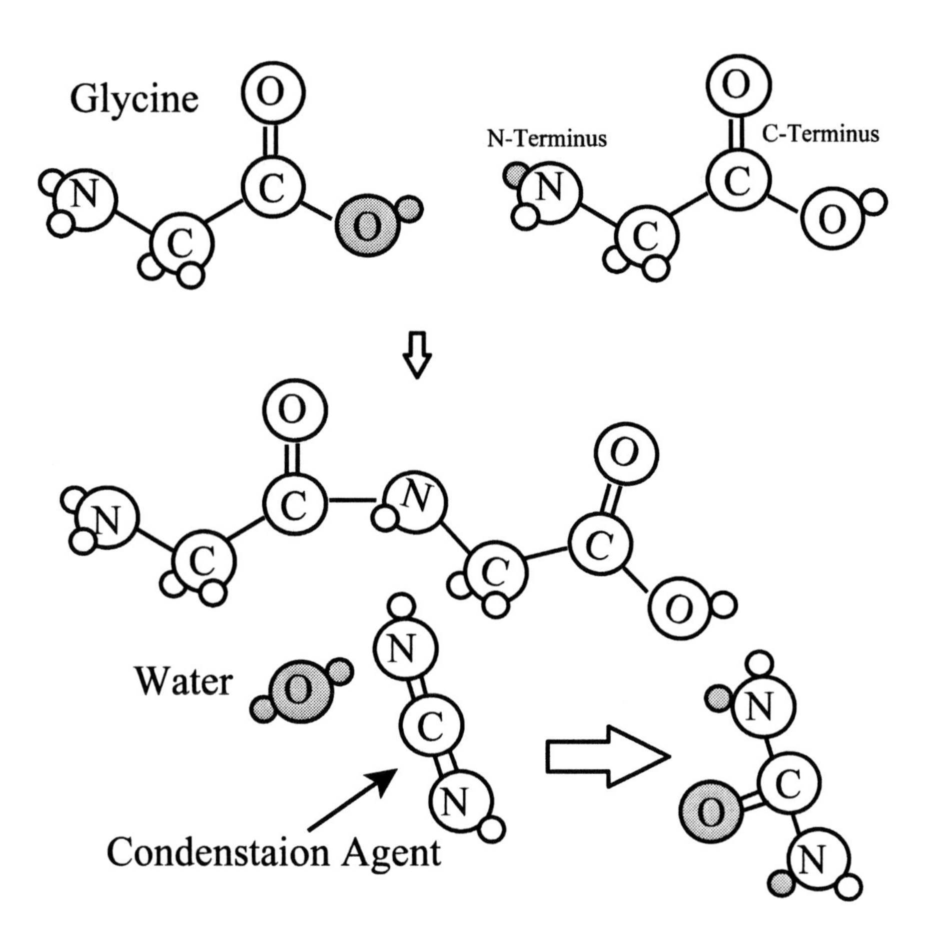

RNA Synthesis

When researchers moved from the living protein theory to the living RNA theory, they unknowingly took a giant step backwards. The decision to switch was made for two reasons: 1) Conceptually, RNA should be able to replicate itself much more easily than a protein and 2) RNA can under special circumstances regulate a few chemical reactions. This second finding solidified RNA as the natural choice for the first living organism. The hope was that it might be able to regulate its own synthesis and thus be a very effective self replicating molecule. Nevertheless, the living RNA theory has created a whole new set of problems that need a solution. Many are much more difficult than the problems created by the living protein theory.

1) The building blocks for RNA are harder to synthesize under plausible prebiotic conditions than amino acids. In fact, cytosine has never been synthesized. Cytosine is also absent from meteorites.

2) Unlike amino acids, two of the building blocks required for RNA (cytosine and ribose) are not stable and have very short lifetimes. It is unlikely that these molecules existed in the soup.

3) Just like proteins, the building blocks for RNA do not form RNA molecules unless water is excluded. Given the short lifetimes of many of the RNA subunits, the high temperatures required to drive off water just accelerate decomposition.

RNA Building Block Synthesis

Creating several amino acids is easy. The hard part is coercing the amino acids to link together in a chain to form a protein. RNA proves much more difficult because even the building blocks are hard to synthesize. Furthermore, once they are created, they do not last long. This makes it difficult to understand how the necessary building blocks achieved a suitable concentration for further reactions. Several key building blocks will now be considered.

Adenine and Cytosine

Adenine has been synthesized in the lab from concentrated solutions of hydrogen cyanide and ammonia. While this process works in the lab, it is not clear how the necessary conditions to create adenine would arise in nature.

To synthesize significant quantities of adenine, a concentrated solution of hydrogen cyanide and ammonia is required. Concentrating hydrogen cyanide and ammonia under plausible conditions is problematic. Hydrogen cyanide is a very reactive chemical. In low concentrations, it reacts with water to form many products that are not adenine. These side reactions use up the hydrogen cyanide and lower its concentration. To make the process more difficult, one of the most abundant chemicals produced in the early atmosphere was undoubtably formaldehyde and "Formaldehyde reacts spontaneously with hydrogen cyanide to form cyanohydrin, a well known reaction that has vexed workers in the field of prebiotic chemistry relying on the unencumbered availability of HCN in high concentration to form a plethora of evolved molecules."[21] Ammonia is equally problematic because it decays rapidly when exposed to sunlight,[15] and it boils at sub-freezing temperatures. So while some adenine might be formed under plausible conditions, very little is produced. The high concentrations of ammonia and hydrogen cyanide required to make adenine do not represent plausible prebiotic conditions.[3]

Because adenine has been found in meteorites, there is evidence that it is produced by nature in space.[3] Nevertheless, based on the above discussion, adenine was certainly a very rare chemical 4 billion years ago.

Cytosine is much more problematic than adenine. It has never been produced under any plausible prebiotic conditions, even in minute quantities. It is not found in meteorites, so it is not easily synthesized in space. Cytosine is not stable in water. Its lifetime depends on the temperature. At 100 degrees Celsius, cytosine decomposes in 19 days. At room temperature, the decomposition is 340 years. These observations have led Miller and several other researchers to suggest that Cytosine was not found in the first self replicating molecule. [5,6]

Ribose

Ribose is the most troublesome subunit. It can only be synthesized in small quantities under plausible prebiotic conditions, and its lifetime in water is extremely short (73 minutes at 100 degrees Celsius, and 44 years at 0 degrees Celsius). Given that it is hard to synthesize in large quantities and that it decays rapidly once it is produced, it is difficult to see how a reasonable concentration of ribose ever existed in the soup. Many scientists including Miller have suggested that the first RNA molecules probably did not include ribose (thus the term pre-RNA).[7]

Ribose presents another difficulty. Just like amino acids, sugars have isomers (mirror images). It has been found experimentally[8,9] that these isomers interfere with self replication. The interference is severe because it terminates the growing chain. So when the first living prebiotic RNA tries to replicate, it must do so in an environment enriched in one isomer of ribose. No mechanism for such an enrichment has been proposed by researchers.

Many scientists have decided that the problems with ribose are so severe that the molecule should be excluded as a possible building block. Since ribose is just the glue that holds RNA together, other chemicals should be able to take its place.[8]

Finally, ribose is a reducing sugar. This means that it will react very quickly with amino acids, and the resulting polymer will fall out of solution. Any ribose in the soup will quickly be eliminated by reactions with the amino acids in the soup.

A Pre-RNA World?

The most recent origin of life theory involves a pre-RNA living molecule. This molecule probably lacked cytosine and ribose. Because such a molecule no longer exists in life, it is hard to address all the possible candidates. How can one possibly test an hypothesis phrased as follows: We believe that some chemical (but we don't know what it was) at one time lived on earth, and this chemical was capable of self replication. We are confident that one day we will find it, and prove our hypothesis correct.

Assembling the Building Blocks

The building blocks for RNA are called nucleotides. A nucleotide consists of 1 phosphate group attached to a ribose which in turn is attached to one of the four bases, uracil, cytosine, adenine or guanine.

In the case of proteins, the amino acid is the smallest building block. No condensation reactions are required to create amino acids. In contrast, two condensation reactions are required to create a nucleotide. One to attach the phosphate to ribose and one to attach one of the bases (adenine, guanine, uracil, and cytosine) to the ribose (figure 8.10).

Under optimal conditions, adenine and guanine can be attached to ribose in the lab. The procedure involves dry heat and sea water. Nucleotides that use cytosine and uracil have no plausible mechanisms to attach the base to ribose.[10]

Figure 9.8: Condensation Reaction (Subunits of RNA)

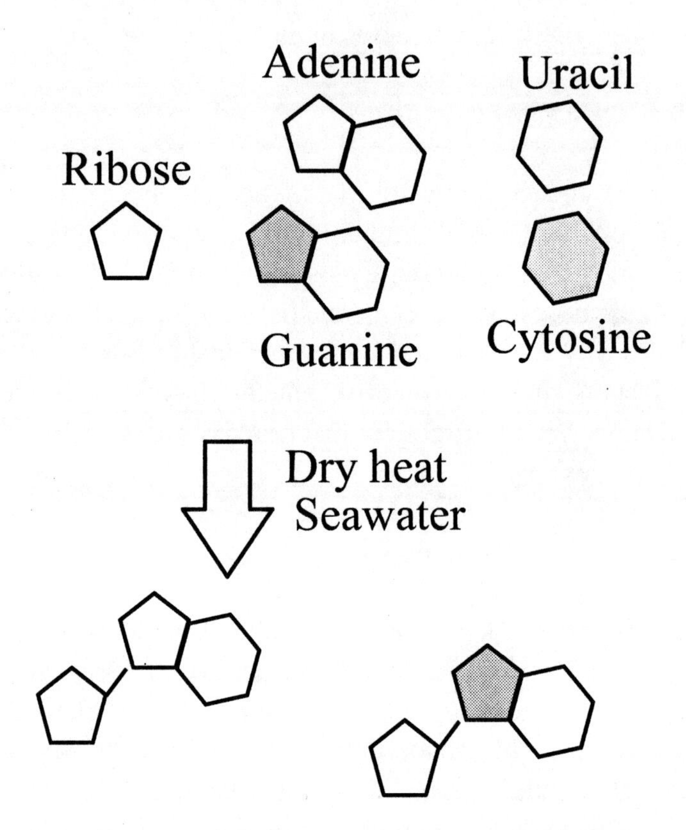

Activated Monomers

Condensing agents are popular for peptide synthesis. They are also effective for RNA synthesis, but in the case of RNA, the nucleotides are usually directly activated before being added to the mixture. In the presence of clay, such activated monomers have been shown to form chains greater than 50 nucleotides long.[16] The most popular activation agent is impA and is shown in figure 9.9. To form impG, impC, and impU replace the adenine with the appropriate base. The source of these activated molecules is unknown. They would have not been present in the soup, so when researchers add them to test tubes in high concentration and then claim that their experiment models the origin of the life, their claims are without merit.

Figure 9.9: ImpA

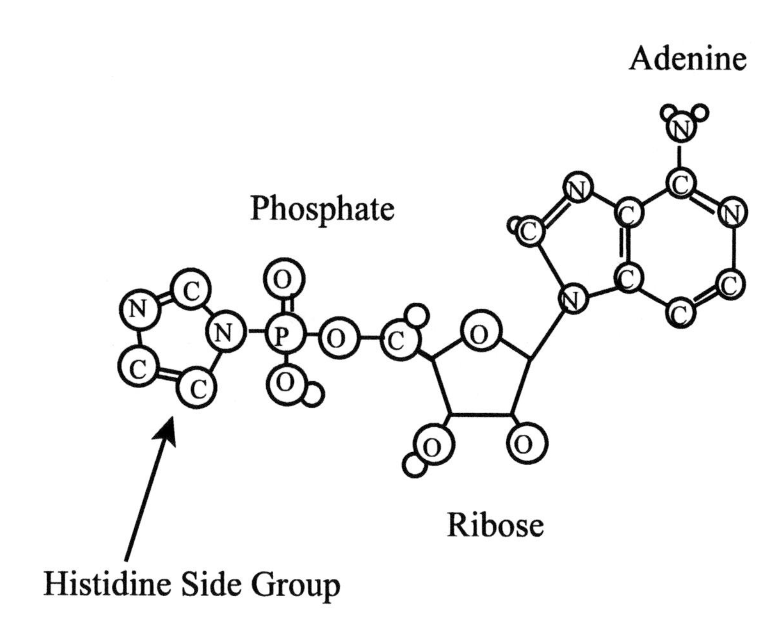

Review of Investigator Interference

Investigator interference will now be summarized. As mentioned earlier, investigators do not have 5 billion years to observe experiments, so some interference is necessary.

Interference Strategy #1: Eliminate the Undesirable Chemicals

If chemical A and chemical B react to form chemical P, then this chemical reaction can be written as A+B—> P. Suppose that Miller's water trap contains 3 chemicals, A, B and C. The possible reactions involving the chemicals are as follows: A + B–>P and A +C->D.

Unfortunately, the second reaction is favored. So after a few days all of the chemicals in the flask are D, but the researcher desires chemical P. So instead of using the contents of the flask to create P, he orders A and B from his chemical supplier. He mixes these two chemicals while applying heat, and the product is P. This process is how organic chemists make chemicals. They control the chemicals that they start with, and this influences the products that they get. Applying this technique to origin of life scenarios is questionable because it is not clear how nature can exclude the undesirable chemicals.

In chapter 7, figure 7.1 shows that one of the functions that enzymes perform is to eliminate undesired reactions. They accomplish this by speeding up the desired reactions. When investigators manipulate the chemicals in their system to create a desired product, they are mimicking this particular mode of enzyme action. They are using their knowledge of chemistry because the required molecular knowledge is not present in the system.

Examples of cross reaction elimination:

- Fox's thermal proteins. He did not include carboxylic acids or other organic components (like aldehydes) that might terminate a growing protein chain.

- The most extreme examples of cross reaction elimination involve RNA. The reason is that ribose is included freely, but no amino acids are included. This is not a plausible condition. Amino acids react very quickly with sugars like ribose to create very long chain polymers. Anyone who has baked cookies or toasted a piece of bread is familiar with this reaction. Browning is caused when amino acids (especially lysine) react with sugar. This reaction would make any sugar present in the primordial soup unavailable for RNA formation.[11,12]

Interference Strategy #2: Concentrating Volatile Chemicals

Concentrated formaldehyde is critical for the synthesis of ribose. Concentrated hydrogen cyanide and ammonia are critical for the synthesis of adenine. It is not clear how these chemicals could ever be present in high concentrations on the early earth.[13] How does one concentrate a chemical that boils at sub-freezing temperatures in a small puddle? This is a difficult problem.

Interference Strategy #3: The Use of Condensation Agents or Activated Monomers

Condensation agents help form many of the bonds that are necessary in biological precursors, whether RNA or protein. Condensation agents remove water and by doing so promote the formation of large biological molecules. Condensation agents were discussed for proteins, but they have also been used to successfully join RNA nucleotides into short chains. RNA synthesis usually just skips this step, and instead researchers usually just add an activated monomer like impA, impG, impC, or impU.

There are no plausible synthesis mechanisms for the condensation agents or the activated monomers. If they are created in Miller's spark experiment or if they exist in meteorites, then the amount present is minuscule. How some investigators can add these chemicals to reactions in massive qunatities, and still think that they are modeling plausible prebiotic conditions is certainly an unsolved mystery.

Nevertheless, the motivation for using these techniques is clear. Without these techniques, the biological precursors are limited to a size that is too small to be biologically active.[16] Given that condensation agents and activated monomers are often coupled with carefully timed washes designed to grow the protein or RNA molecule, the analogy to blowing up the door in figure 9.4 definitely applies.

Interference Strategy #4: Controlling the Energy Sources

In most experiments, destructive energy sources are eliminated by the investigator. For example, if the trap in Miller's spark chamber is illuminated with UV light, many of the products will be destroyed.[4]

Interference Strategy #5: Substituting Human Knowledge

This is the most subtle form of interference, and the most common. In systems that lack the required molecular knowledge, it is very easy for researchers to unintentionally add knowledge to the system through the design of their experiment.

The carefully controlled sequential washes that accompany many RNA and protein chain elongation experiments are a perfect example. Often a growing RNA or protein molecule is attached to a stationary substrate, activated nucleotides or amino acids are added, and a rinse is applied after the desired chemical bond forms. This form of interference is present in most prebiotic experiments, and sometimes it goes unnoticed.

Conclusion:

The goal of this chapter was to show that the precursors to life whether RNA or proteins are extremely difficult to create. Maybe one or two such molecules are expected given optimal conditions and 5 billion years. The design inference based on this conclusion alone is very strong. The inference will be strengthened in the next chapter. The next chapter will show that the knowledge required for self replication is very large. If the entire ocean is packed tight with either proteins or RNA, then the odds that one of the molecules can self replicate is still zero. Several thousand bits of knowledge are required, and zero tries (or almost zero) will never allow chance to find a solution.

Many investigators researching the origin of life are disappointed with their progress, and this shows in the scientific literature. Today, it is acceptable to publish an article that is critical of the origin of life paradigm as such articles do get published.

Any publication suggesting the possibility of design is either rejected or starts a witch hunt in which the editor who approves the article is the target. The first step in any scientific revolution is to realize that there is a problem with the current theory, and for many scientists this realization has already taken place. Joyce and Orgel summarize the situation as follows:

"In our initial discussion of the RNA World we will accept The Molecular Biologist's Dream: "Once upon a time there was a prebiotic pool of Beta-D-nucelotides . . . . We will now consider what would have to happen to make the dream come true. This discussion triggers the Prebiotic Chemist's Nightmare: how to make any kind of self replication system from the intractable mixtures that are formed in the experiments designed to simulate the chemistry of the primitive earth."[20]

References:

1) Temussi, et al., "Structural Characterization of Thermal Prebiotic Polypeptides," Journal of Molecular Evolution, p105-110, 1976.
2) Miller, Orgel, The Origins of Life on Earth, Prentice Hall, 1974.
3) Shapiro, "The Prebiotic Role of Adenine: A Critical Analyis," Origins of Life and the Evolution of the Biosphere," 25:83-98, 1995.
4) Thaxton, Bradley, Olsen, The Mystery of Life's Origin: Reassessing Current Theories, Philosophical Library, 1984.
5) Levy, Miller, "The Stability of the RNA bases: Implications for the Origin of Life," PNAS, 95: 7933-7937, 1998.
6) Shapiro, "Prebiotic Cytosine Synthesis: A Critical Analysis and Implications for the Origin of Life," PNAS, 96: 4396-4401, 1999.
7) Larralde, Robertson, Miller, "Rates of decomposition of Ribose and other Sugars: Implications for chemical Evolution," PNAS, 92: 8158-8160, 1995.
8) Joyce, Schwartz, Miller, Orgel, "The Case for an Ancestral Genetic System Involving Simple Analogues of the Nucleotides," PNAS, 84: 4398-4401.1989.
9) Joyce, Visser, Boeckel, Boom, Orgel, Westrenen "Chiral Selection in Poly (C) Directed Synthesis of Oligo (G)," Letters to Nature, 310: 602-604,1984.
10) Fuller, Sanchez, Orgel, "Studies in Prebiotic Synthesis. V11 Solid State Synthsis of Purine Nucleosides," Journal of Molecular Evolution, 1:249-257, 1972.
11) Nissenbaum, Kenyon, Oro, "On the Possible Role of Organic Melanoidin Polymers as Matrices for Prebiotic Activity," Journal of Molecular Evolution. 6:253-270, 1975.
12) Thaxton, Bradley, Olsen, The Mystery of Life's Origin: Reassessing Current Theories, Philosophical Library, pp 60-61,1984.
13) Thaxton, Bradley, Olsen, The Mystery of Life's Origin: Reassessing Current Theories, Philosophical Library, p 64,1984.
14) Ferris, "Prebiotic Synthesis: Problems and Challenges," Cold Spring Harbor on Quantitative Biology, Vol L11: 29-34, 1987.
15) Thaxton, Bradley, Olsen, The Mystery of Life's Origin: Reassessing Current Theories, Philosophical Library, pp 43-44,1984.
16) Ferris, "Montmorillonite Catalysis of 30-50 Mer Oligonucleotides: Laboratory Demonstartion of the Potential Steps in the Origins of the RNA world," Origins of Life and Evolution of the Biosphere, 32:311-332, 2002.
17) Osterberg, Orgel, Lohrmann, "Further Studies of Urea Catalyzed Phosphorylation Reactions," Journal of Molecular Evolution, 2:231-234, 1973.
18) Fox, Dose, Molecular Evolution and the origin of Life, Freeman and Co., 1972.
19) Thaxton, Bradley, Olsen, The Mystery of Life's Origin: Reassessing Current Theories, Philosophical Library, pp 66,1984.

References (continued)

20) Joyce and Orgel, The RNA World, Gesteland, Cech, Atkins, Cold Spring Harbor, "Prospects for Understanding the Origins of the RNA World," p50, 1999.
21) Mojzsis, Krishnamurthy, Arrhenius, The RNA World, Gesteland, Cech, Atkins, Cold Spring Harbor, "Constraints on Molecular Evolution," p20-21, 1999.
21) Fox, Dose, Molecular Evolution and the origin of Life, Freeman and Company, p37, 1972.
22) Shapiro, Origins: A Skeptics Guide to the Creation of Life on Earth, 1986.
23) Overman, A Case Against Accident and Self-Organization, 1997.
24) Plankensteiner, Reiner, and Rode, Stereoselective Differentiation in the Salt-induced Peptide Formation Reaction and Its Relevance for the Origin of life, Peptides, 2004.
25) Bujdak, Eder, Yongyai, Faybikova, and Rode, Investigation on the Mechanism of Peptide Chain Prolongation on Montmorillonite, Journal of Inorganic Biochemistry, 1996.

Readers who wish to read more about chemical evolution and its problems should try to find Thaxton's book, The Mystery of Life's Origin: Reassessing Current Theories, in their local university library. Unfortunately, the book is out of print. His book was the primary reference for this chapter. The two papers by Shapiro, reference 3 and 6, are also excellent resources.

Chapter 10: Self Replicating Molecules and Systems

This chapter will attempt to quantify the amount of molecular knowledge needed for self replication. Both proteins and RNA will be considered. While many researchers have theorized that one of these molecules emerged as the first self replicator, origin theories stand a much better chance if both are involved. While RNA can perform some of the functions normally performed by proteins, proteins are much more efficient. Amino acids have many functional groups available in their side chains, and these functional groups impart to proteins a versatility than RNA cannot possibly possess. To understand why a system comprised of both is better, consider how numbers and letters are used in the following two sentences.

- The number is 4,900,555,015 dollars.
- The number is four billion nine hundred million five hundred fifty five thousand and fifteen dollars.

Often numbers communicate numerical concepts better than words. The first sentence is much easier to understand. Forcing RNA to do the job of a protein is clumsy. It is analogous to writing out a very large number using words to represent the numbers. Just because it is possible, does not mean that it is the easiest or best way to accomplish the task. RNA is good at storing information. Proteins are good at regulating chemical reactions. The first system of replicating molecules was probably a combination of both, and a good model for such a system is alive and well today in the simplest bacteria. Nevertheless, because chemical evolution does not explain the spontaneous emergence of bacteria from the primordial soup something simpler needs to be considered. The goal of this chapter is to show that something simpler does not work because simple systems cannot self replicate.

A Self Replicating Peptide

In 1996, an article was published in Nature in which David Lee reports to have found a self replicating peptide.[1] The title of the article is appropriately "A Self Replicating Peptide." Unfortunately, the investigator interference required for self replication is perhaps the most extreme in the history of origins research.

The peptide of interest contains 32 amino acids. The sequence is as follows:

arg-met-lys-gln-lys-glu-glu-lys-val-tyr-glu-lys-lys-ser-lys-val-ala-cys-leu-glu-tyr-glu-val-ala-arg-leu-lys-lys-leu-val-gly-glu.

The peptide does not self replicate using amino acids. Instead it uses a pool of two peptides, one is 17 amino acids long and the other is 15 amino acids long. The amino acid sequences of these two peptides are shown below. Notice that if a peptide bond forms between ala (last amino acid on right in the peptide with 17 amino acids) and cys (first amino acid on left in the peptide with 15 amino acids) then a replica of the self replicating peptide results.

arg-met-lys-gln-lys-glu-glu-lys-val-tyr-glu-lys-lys-ser-lys-val-ala

cys-leu-glu-tyr-glu-val-ala-arg-leu-lys-lys-leu-val-gly-glu

Because the peptide with 32 amino acids facilitates the formation of this single peptide bond, Lee claims that this peptide can self replicate. But is this really true? To self replicate, this peptide requires a pool of two peptides. One of these peptides has the same amino acid sequence as the first 15 amino acids in the self replicating peptide, and the other has the same amino acid sequence as the next 17 amino acids. Where do these peptides come from? In this case, they are supplied by the investigator.

Chapter 9 discussed the difficulties of creating peptide chains under plausible prebiotic conditions. Due to the difficulties, peptides with more than six amino acids are expected to be very rare chemicals. Peptides composed of 15 to 17 amino acids will be much more scarce. Yet to self replicate, this peptide requires an abundant supply of both, and not just any peptide. One of these peptides must be identical to the first half of the self replicator, and the other peptide must be identical the second half of the self replicator.

This last requirement is particularly troublesome. Suppose the self replicator comes into contact with two random peptide chains. One is 15 amino acids long and the other is 17. How often will the two smaller peptides be an exact replica of the self replicator? Answer 1 time in every $4x10^{41}$ tries (assuming that every amino acid has a 1 in 20 chance of occurring at each position). Given the low concentration of peptides in the primordial soup, the probability for such an encounter is zero.

The interference does not stop here. It is critical that the first amino acid in the peptide with 15 amino acids be a cysteine. Cysteine has chemical properties that facilitate peptide bond formation, and to make sure that the interference sets the record, the alanine (last amino acid on right in the peptide with 17 amino acids) must be chemically altered to make it much more susceptible to attack by the sulfur atom in cysteine's side chain.

Finally, the self replicating peptide contains eight lysines. Lysine is instrumental in its self replication as its charge plays a role in aligning the two small peptides. Lysine is one of the amino acids that has yet to be synthesized under plausible prebiotic conditions. So even if lysine was present in the soup, its concentration would have been negligible.

Every possible strategy of interference is employed by this investigator to promote replication. This mixture of peptides has almost no chance of existing on the primitive earth. Even if it did, as soon as the supply of 15 and 17 amino acid peptides runs out, the replication stops. Despite all of this interference, the claim of self replication is not valid. Self replication involves a system that can duplicate all of its components. In this system, the self replicating peptide is supplied with one peptide containing 15 amino acids and one with 17 amino acids. A true self replicating molecule could generate these two smaller peptides from the amino acids in the primordial soup.

The authors of this paper tried to use dynamite to blow up the door in figure 9.4, but the door withstood the blast and did not open. So the authors just claimed that it opened.

Proteins do not self replicate, and this explains why most scientists rejected the self replicating protein hypothesis in favor of the self replicating RNA hypothesis.

RNA Self Replication

Conceptually RNA should be able to self replicate without the help of proteins. This is shown in figure 10.1. The original strand serves as a template. New base pairs arrive and form weak bonds with their complement. Adenine can form a bond with Uracil, and Guanine can form a bond with Cytosine. After one replication, two complementary strands exist. Another round of replication is necessary to duplicate the original strand. The complement to the original strand is also free to make more copies.

Figure 10.1: Conceptual Model for RNA Self Replication

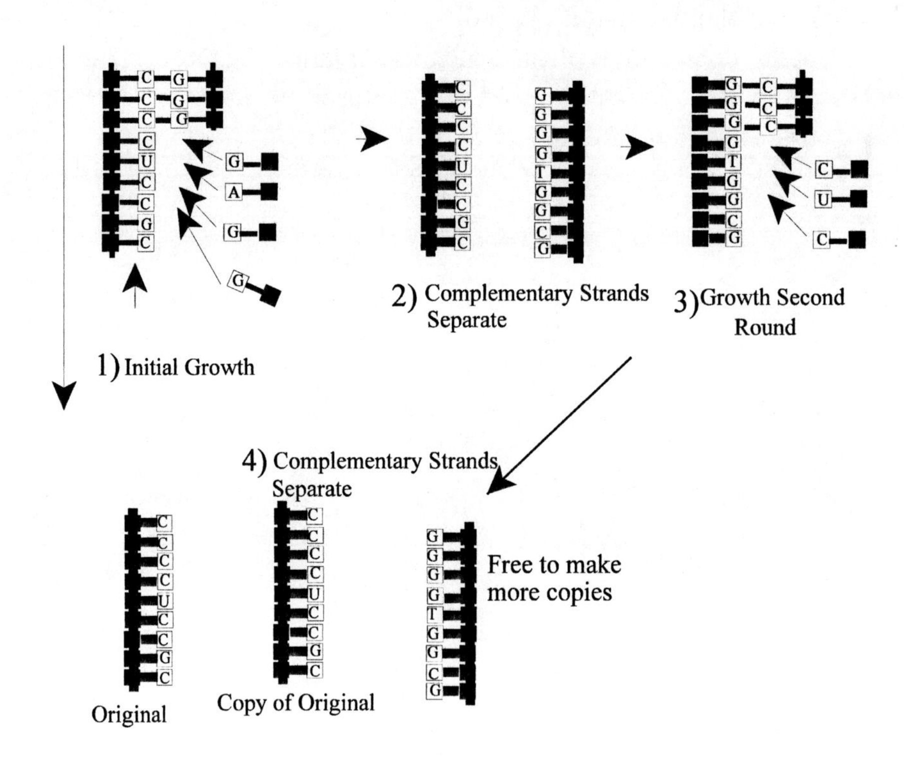

On paper, this model is great. Nevertheless, it does not work in the lab. The problems were described by Joyce and Orgel as follows:

- Most strands of RNA are unsuitable templates. The original RNA molecule that serves as the template must contain a very high concentration of cytosine to make process 1 in figure 10.1 viable.[2,3] This situation is unlikely to be met because as discussed earlier cytosine has no plausible prebiotic synthesis pathway and it decays rapidly. Nevertheless, the original strand depicted in figure 10.1 meets the high C requirement.

- The chain will not grow correctly unless a very specific activation agent is used to activate the nucleotides. The activation agent of choice is not ATP (GTP, CTP or UTP). While life uses these, if these activation agents are used without proteins the phosphate groups usually attach to the wrong carbon atom in ribose.[2,3] ImpA, impG, impU and impC are the activation agents of choice. These activation agents contain the same side group as the amino acid histidine, which is one of the three amino acids that have not been synthesized in prebiotic experiments. Thus, it is unlikely that these activation agents were present in the primordial soup.

- The complement of the original chain will have a high G content. This is inevitable due to the requirement for high C in the original chain. This is problematic because RNA with a high concentration of guanine tends to fold up in such a way that it cannot be an effective template for replication.[2,3] Thus, the second round of growth in figure 10.1 does not happen.

- If different isomers of ribose are present, these isomers will terminate the growing chain.[2,3]

Joyce and Orgel comment that "In light of the available evidence, it seems unlikely that a pair of complementary sequences can be found each of which facilitates the synthesis of the other . . ." [3]

Just to add to the difficulties, if too many steps form in the replication ladder (complementary bonds between base pairs), then the strands will never separate.[4] Furthermore, figure 10.1 is oversimplified in that it does not show that in order for the RNA strands to grow, an RNA enzyme is required to catalyze the reaction. Because a growing chain cannot catalyze its own replication, two identical RNA molecules must arise simultaneously in the soup. Each capable of replicating the other.

A pattern is beginning to emerge for the RNA world. The RNA world is a speculative world without proteins where RNA is the most important molecule. RNA regulates all chemical reactions and contains all of the molecular knowledge for life. The pattern that is emerging is that perhaps this world is too speculative in that it may have never existed.

Again Joyce and Orgel put it best: "Scientists interested in the origins of life seem to be divided neatly into two classes. The first, usually but not always molecular biologists, believe that RNA must have been the first replicating molecule and that chemists are exaggerating the difficulties of nucleotide synthesis . . . The second group of scientists are much more pessimistic. They believe that the de nova appearance of oligonucleotides on the primitive earth would have been a near miracle. The author's subscribe to this latter view. Time will tell which is correct." [3]

One last point, RNA replication in the lab makes use of extensive investigator interference. Chemicals like amino acids, aldehydes, and sugars (other than ribose) are arbitrarily excluded. Very specific activation agents are used to encourage replication (ImpA for adenine, ImpG for guanine, ImpC for cytosine, and ImpU for uracil). The concentration of the chemicals (especially cytosine and ribose) is billions and billions of orders of magnitude higher than what one would expect under plausible prebiotic conditions.

Dynamite is being used to blow the door open in figure 9.4, and the door is just too solid. It remains closed and the scientist remains trapped. Fortunately, many scientists understand this, and they no longer claim that the door is open.

How Much Knowledge is Required to Create a Ribozyme

RNA molecules capable of facilitating chemical reactions do exist. Because such RNA molecules perform a role traditionally carried out only by protein enzymes, they are called ribozymes. Ribozymes have been shown to facilitate the creation of both peptide bonds in proteins, and the bonds between phosphate and ribose in RNA. This discovery is very significant in that it means RNA can both store and implement knowledge. It also explains the popularity of RNA as the first living molecule.

Bartel carried out a very relevant experiment. In this experiment. 65 ribozymes were isolated from a pool of $1x10^{15}$ RNA molecules. All ribozymes isolated contained 200 bases. This result allows for a direct calculation of the knowledge in ribozymes. If 65 sequences have some minimal enzymatic activity out of a pool containing 10^{15} random sequences, then one in every 15 trillion sequences is a ribozyme. Thus the molecular knowledge is as follows: knowledge = 3.32 x log (15 trillion) or 44 bits. Note that knowledge and not information is used because the 65 ribozymes were not yet optimized. The experiment also subjected the ribozymes to several rounds of selection in which only the best were chosen. Selection dramatically improved their catalytic efficiency.

Given the extreme difficulties associated with synthesizing an RNA molecule containing 200 or more bases, it is unlikely that even one such molecule ever existed on the primitive earth, and 15 trillion are needed to just get 65 functional ribozymes. Furthermore, ribozymes are not self replicators. The knowledge required for self replication is certainly many orders of magnitude more than the 44 bits required for a marginally functional ribozyme. Finally, the 44 bits calculated above are for evolution in a test tube where all competing side reactions are eliminated. If the primordial soup contains free amino acids, aldehydes, and undesirable isomers of ribose, then the 44 bits will increase by a factor of at least 10 and probably more.

Molecular Knowledge in the Primordial Soup

In chapter 5, the difficulties with creating a functional protein in the primordial soup were explored. A similar analysis will now be undertaken for RNA. Because of the scarcity of the RNA subunits (especially ribose and cytosine), the information content of any RNA molecule that evolves in the soup is expected to be very high.

If the soup existed, its exact composition is unknown. Nevertheless, several generalizations are possible. Ribose and cytosine should be extremely rare (see chapter 9). Furthermore, ribose will react with any free amino acids in the soup forming an insoluble polymer. Adenine can be synthesized in the lab, but not under plausible conditions with high yield. Even phosphate will be scarce if inorganic salt is present in the soup.[7]

While the concentration of cytosine and ribose in the soup is probably zero, applying information theory to this situation is not productive because infinite information, implies zero chance for success. So instead this section will make some very favorable assumptions concerning the composition of the soup. The assumptions are not realistic. They are made for educational purposes only.

Favorable Assumptions:

1) All phosphate, sugar and base molecules in the soup exist only as activated nucleotides. That is any adenine in the soup is assumed to be attached to ribose or another sugar. All sugars either have a high energy phosphate group attached or they are attached to some other activating agent.

2) No amino acids are found in the soup. While these are easily synthesized in prebiotic experiments, they must be excluded as they react quickly with ribose and other aldehydes, removing ribose from solution and preventing more ribose from forming. Amino acids and ribose cannot coexist in the soup.

3) No aldehydes exist in the soup. While these are required for the synthesis of ribose and other sugars, they cannot be allowed to persist. Aldehydes react with the four biological bases. These reactions will interfere with the formation of RNA.

Given this starting point, what is the probability that an RNA molecule will emerge from the soup? Assume the following:

- Every time an activated nucleotide attacks a ribose, it has a 50% chance of attacking the wrong carbon atom. This results in premature chain termination.[2,3]

- Half of the ribose present is the wrong isomer, this also results in premature chain termination.[2,3]

- 3/4 of the bases attached to the ribose are not biological. That is adenine, guanine, cytosine, and uracil are only used in 1/4 of the activated nucleotides. The most common base is likely ammonia or some other simple amine.

- 3/4 of the activated nucleotides use a sugar other than ribose or deoxyribose. This also results in premature chain termination. Given that ribose is usually only a minor product in any prebiotic experiment that synthesizes simple sugars, this is a very generous assumption.

Even with these most favorable assumptions that ignore all competing side reactions, every nucleotide added to the RNA chain still contributes a minimum of 6 bits of primordial information (for every 64 nucleotides added to the chain, only 1 is expected to be biologically relevant, and this corresponds to 3.32 x log (64/1) = 6 bits of primordial information). This is three times the value calculated for amino acids in chapter 5.

Thus, a 200 base pair random RNA sequence contains 6 x 200 = 1200 bits of primordial information, and as explained in chapter 5, primordial information can be related to a probability because it is a form of knowledge - the knowledge to exclude chemicals found in the soup that are not used by life today.

Thus, a 200 base pair random RNA sequence has a 1 in 2^{1200} chance of emerging in the primordial soup. Given that only 65 out of 15 trillion will exhibit any ribozyme functionality, the odds are staggering - 1 time in 3.9×10^{372} tries. Furthermore, this calculation is only for a ribozyme capable of regulating a simple chemical reaction. The odds of a self replicating ribozyme emerging are certainly much smaller.

In summary, the probability of creating a 200 base ribozyme is extremely small because so few random sequences contain the required knowledge, but given that no 200 base RNA molecules existed on the primitive earth, the odds are no longer almost zero, but instead almost zero multiplied by zero.

Finally, as noted in chapter 5, using information theory to calculate the odds has some drawbacks. Information theory only takes into account the concentration of the various chemicals. It does not have the ability to deal with chemical properties that may make certain reactions more probably, and this can skew the results in favor of evolution or against it. In the case of RNA, a very strong argument can be made that the skewing is strongly in favor of evolution. This is because the above calculation excluded amino acids and aldehydes from the soup. Thus, the information calculated above represents RNA that evolves in a test tube, not the real world.

Self Replication and Perpetual Motion

Researchers today are actively seeking and finding new ribozymes. Many are artificially engineered and others arise from random sequences. Many of these researchers believe that in time they will find a self replicating RNA molecule. Others like Joyce and Orgel who are at the forefront of the research disagree.

In chapter 7, several techniques used by life to circumvent the second law of thermodynamics were discussed. Unless a self replicating RNA molecule has the capability to implement some of these same techniques, its existence can be ruled out on purely theoretical grounds.

Based on fundamental laws of physics, science can state with certainty that if a self replicating RNA molecule is found, the molecule will only be able to replicate in a test tube. It will require a continuous supply of activated nucleotides to drive its replication. While this might work in the test tube, it would certainly not work in the primordial soup. Activated nucleotides in the soup would not last for more than a few days. Since their decay would dominate any conceivable path for prebiotic synthesis, activated nucleotides in the soup would be very rare and probably non-existent.

Given the difficulty associated with the prebiotic synthesis of ribose, adenine, and cytosine, the concentration of these critical molecules in the soup would be extremely low (probably negligible). This means that the first successful self replicating RNA molecule must be able to direct the synthesis of adenine, cytosine, ribose, uracil and guanine. If it cannot do this, it will not be able to replicate in the soup. Furthermore, it must be able to activate all of these nucleotides. So this special RNA molecule must know how to tap a plentiful energy source and use it to drive many different chemical reactions. If it cannot perform all of these functions, then it is a perpetual motion machine, and its very existence is limited to biology textbooks.

Figure 10.2: A Self Replicating RNA Molecule

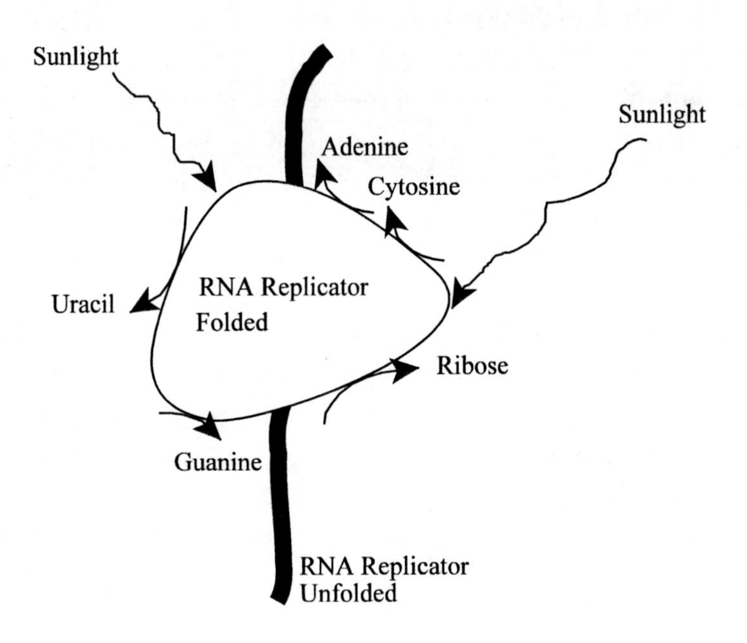

In figure 10.2, the RNA molecule can exist in two states, folded and unfolded. When folded, it catalyzed RNA replication, and the formation of adenine, ribose, cytosine, uracil, and guanine. It also must create activated nucleotides (not shown). When unfolded, it serves as a template for replication. The folded version must also know how to replicate the unfolded version.

This particular ribozyme taps into sunlight as an energy source using a primitive form of photosynthesis. Other self replicating RNA molecules could potentially oxidize a chemical like methane, hydrogen, or sulfur to generate the required energy.

Figure 10.2 is what is required of a "living molecule." Anything less is not alive. This figure was constructed with due consideration to the second law. Any RNA molecule that does not possess all of the capabilities shown in figure 10.2 is a perpetual motion machine. It may replicate in the lab as long as it is supplied with activated nucleotides, but it will not replicate in the soup. Thus, it only exists in textbooks, and there is no need to wait to see if researchers can locate it.

Inventors have been trying to invent perpetual motion machines for at least 2000 years. They have all failed. Nevertheless, many have been issued patents by various governments throughout the world. Two examples of perpetual motion are shown in figure 10.3. Both examples are equally absurd. While many scientists apparently only recognize the absurdity of the first picture, nature can recognize both, and it does not allow either to exist.

Figure 10.3: Perpetual Motion Machines

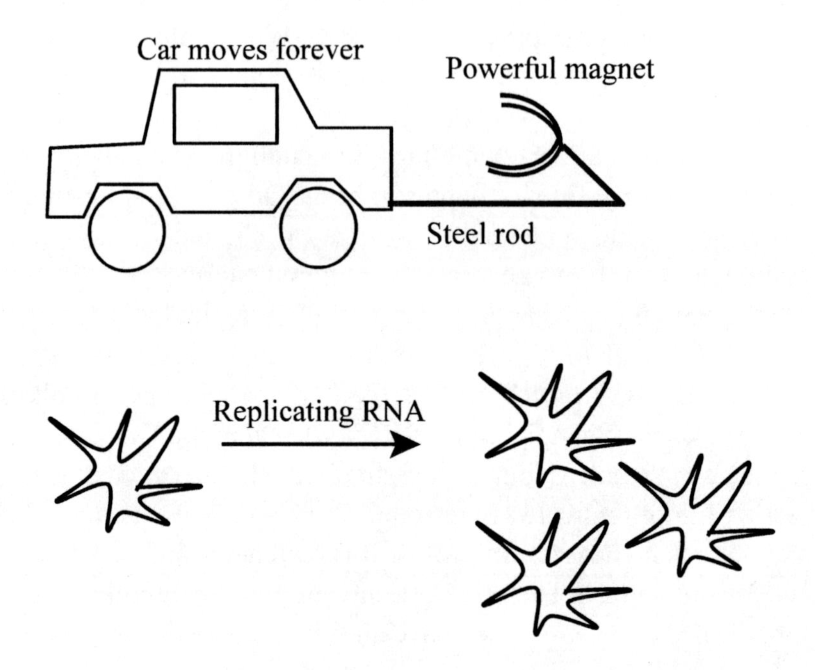

The first picture in figure 10.3 is a clear violation of energy conservation. It does not work because the force that the magnet exerts on the car is exactly cancelled by the force that the car exerts on the magnet. The magnet does not cause the car to move. The second violation is more subtle only because it violates a different law of nature. When a self replicating molecule replicates, the replication decreases the entropy of the universe. The second law is violated. To get around this problem, any real self replicator must know how and be able to couple its replication to a plentiful energy source. If it is unable to do this, then it is a special type of perpetual motion machine, and it only exists on paper and in the imagination of researchers.

References:

1) Lee, Granja, Martinez, Severin, Ghadiri, "A Self Replicating Peptide," Letters to Nature, 382:525-528, 1996.

2) Joyce, Visser, Boeckel, Boom, Orgel, Westrenen, "Chiral Selection in Poly (C) Directed Synthesis of Oligo (G)," Letters to Nature, 310: 602-604, 1984.

3) Joyce and Orgel, The RNA World, Gesteland, Cech, Atkins, Cold Spring Harbor, "Origin of the RNA World," 1999.

4) Bartel, The RNA World, Gesteland, Cech, Atkins, Cold Spring Harbor, "Recreating an RNA Replicase," 1999.

5) Ekland, Szostak, Bartel, "Structurally Complex and Highly Active RNA Ligase Derived from Random RNA Sequences, " Science, 1995.

6) Bartel and Szostak, "Isolation of New Ribozymes from a Large Pool of Random Sequences," Science, 261:1411-1418, 1993.

7) Thaxton, Bradley, Olsen, The Mystery of Life's Origin: Reassessing Current Theories, Philosophical Libraries, 1984.

8) Orgel, Self-organizing Biochemical Cycles, Salk Institute of Biological Studies, 99:12503-12507, 2000.

9) Green, Szostak, "Selection of Ribozyme that Functions as a Superior Template in Self Copying Reaction," Science, 258:1910-1915, 1992.

Chapter 11: The Myth of the Primordial Soup

Any discussion on the origin is life is not complete without considering the primordial soup. There is no direct evidence of the soup's existence, and on purely theoretical ground it should not exist. If it did exist, science can say with certainty that it was a very localized existence. That is it may have been a small puddle, near a volcano, right at the entrance of a cave, near an ocean or a river. The primitive ocean was definitely not the primordial soup. The ocean could not possibly serve as the soup because it would dilute the biological precursors, and it would not protect the precursors from ultraviolet light.

Many authors have criticized the concept of the soup. Its resilience in biology text books is quite amazing given that so few scientists believe that it ever existed.

"Accordingly, Abelson(1966), Hull(1960), Sillen(1965), and many others have criticized the hypothesis that the primitive ocean, unlike the contemporary ocean, was a "thick soup" containing all of the micromolecules required for the next stage of molecular evolution. The concept of a primitive "thick soup" or "primordial broth" is one of the most persistent ideas at the same time that is most strongly contraindicated by thermodynamic reasoning and by lack of experimental support." - Sidney Fox, Klaus Dose on page 37 in Molecular Evolution and the Origin of Life.

"the primitive ocean was steadily irradiated with a relatively high dose of solar ultraviolet light . . . A steady irradiation of a rather homogeneous solution results in degradative rather than synthetic reactions" Sidney Fox, Klaus Dose in Molecular Evolution and the Origin of Life.

"Based on the foregoing geochemical assessment, we conclude that both in the atmosphere and in the various water basins of the primitive earth, many destructive interactions would have so vastly diminished, if not altogether consumed, essential precursor chemicals, that chemical evolution rates would have been negligible. The soup would have been too dilute for polymerization to occur. Even local ponds for concentrating soup ingredients would have met with the same problem. Furthermore, no geological evidence indicates an organic soup, even a small organic pond, ever existed on this planet. It is becoming clear that however life began on earth, the usual conceived notion that life emerged from an oceanic soup of organic chemicals is a most implausible hypothesis. We may therefore with fairness call this scenario the myth of the prebiotic soup." - Thaxton, Bradley, Olsen on page 66 of The Mystery of Life's Origin.

"Contrary to earlier suggestions that essentially all stages of chemical evolution occurred in the open seas, it is now generally accepted that the concentration of the soup was probably too small for efficient synthesis......"- Nissenbaum, Kenyon, Oro, in the "Journal of Molecular Evolution," 1975.

Furthermore, any organic compounds not destroyed by UV light would react to form an insoluble polymer. This reaction known as the Maillard reaction would remove most of the organic molecules in the soup making them unavailable for chemical evolution.

" The rapid formation of this insoluble polymeric material would have removed the bulk of the dissolved organic carbon from the primitive oceans and would thus have prevented the formation of the organic soup." - Nissenbaum, Kenyon, Oro, Journal of Molecular Evolution, 1975.

In summary: 1) It is extremely difficult to create information and knowledge before life exists. 2) Excessive investigator interference is required to make biological subunits polymerize. 3) The prebiotic synthesis of the subunits required for DNA and RNA (especially ribose and cytosine) present some very serious design challenges. 4) It is unlikely that any single chemical can possess the required knowledge to replicate because it must not only know how to replicate, but it must also know how to use an energy source to drive its own replication. 5) Any favorable environment for chemical evolution would have been highly localized to a small puddle. 6) Because of the localized nature of the soup and the low concentration of biological precursors, any robust self replicating system (i.e. Life) would need the ability to synthesize many of the chemicals required for self replication. Any self replicating system lacking this capability would not be able to survive much less replicate.

Taken together the evidence suggests that the first living thing was not a self replicating molecule, but rather a system of chemicals that contained the knowledge required to replicate and the ability to couple this replication to an energy source. Furthermore, the scarcity of chemicals like ribose, adenine, and cytosine imply that for this system to survive, it must have been able to synthesize many if not all such chemicals from more abundant chemicals. All of these factors imply that the first living thing was not that much simpler than life as it exists today. It may have even been more complex.

Part 3: The Evolution of the First Genes

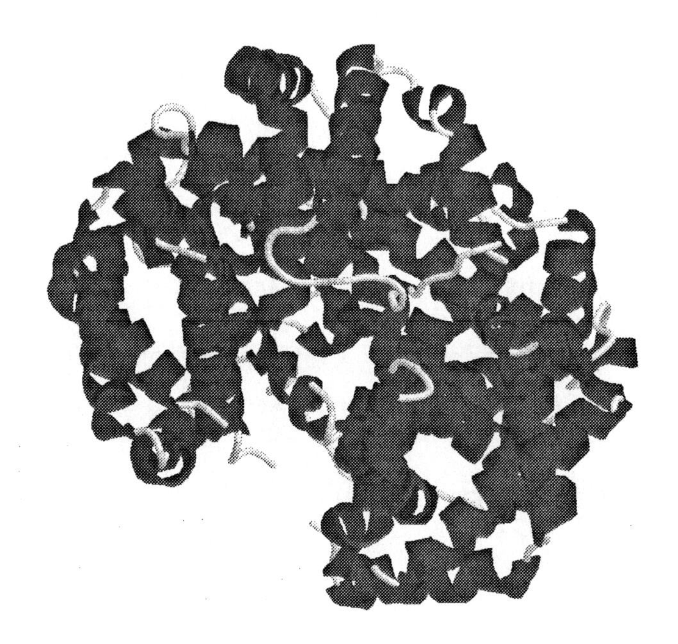

Hemoglobin is the protein that caries oxygen in the blood of many animals.

Chapter 12: Irreducible Complexity

Irreducible complexity was coined by Michael Behe in his book Darwin's Black Box. The idea is that many biological systems require several parts to function properly. Without all of the parts, the system does not provide any selective advantage, and natural selection cannot preserve or optimize it. Behe used a mousetrap to illustrate this concept in his book. The mousetrap does not function properly if any of the pieces are missing. The drawback with the mousetrap analogy is that it does not lend itself to a mathematical analysis.

The trapped scientist example will now be used to show how and why irreducible complexity makes evolution very difficult. Unlike previous examples, the scientist is in a three story building. There are three doors on each floor. But to get to these three doors, the scientist has to open one door that is three stories high (figure 12.1). This door has a very long combination, but once it is open, the combinations for the small doors are easy to find.

Figure 12.1 represents any system that requires three components to function. No selective advantage is realized unless all three components exist. Thus, the first door's combination represents three infons (instead of one). When the first door is opened, all three components exist, and the system conveys a selective advantage. If the scientist ever opens the door, then the knowledge is preserved by the computer. The computer will lock the 27 words needed to open the door, and the scientist will only be able to change 9 words. As he continues, the other doors will be opened very quickly as the steps are small (one word). These small doors optimize the three new genes and their associated proteins.

Figure 12.1: An Irreducibly Complex System

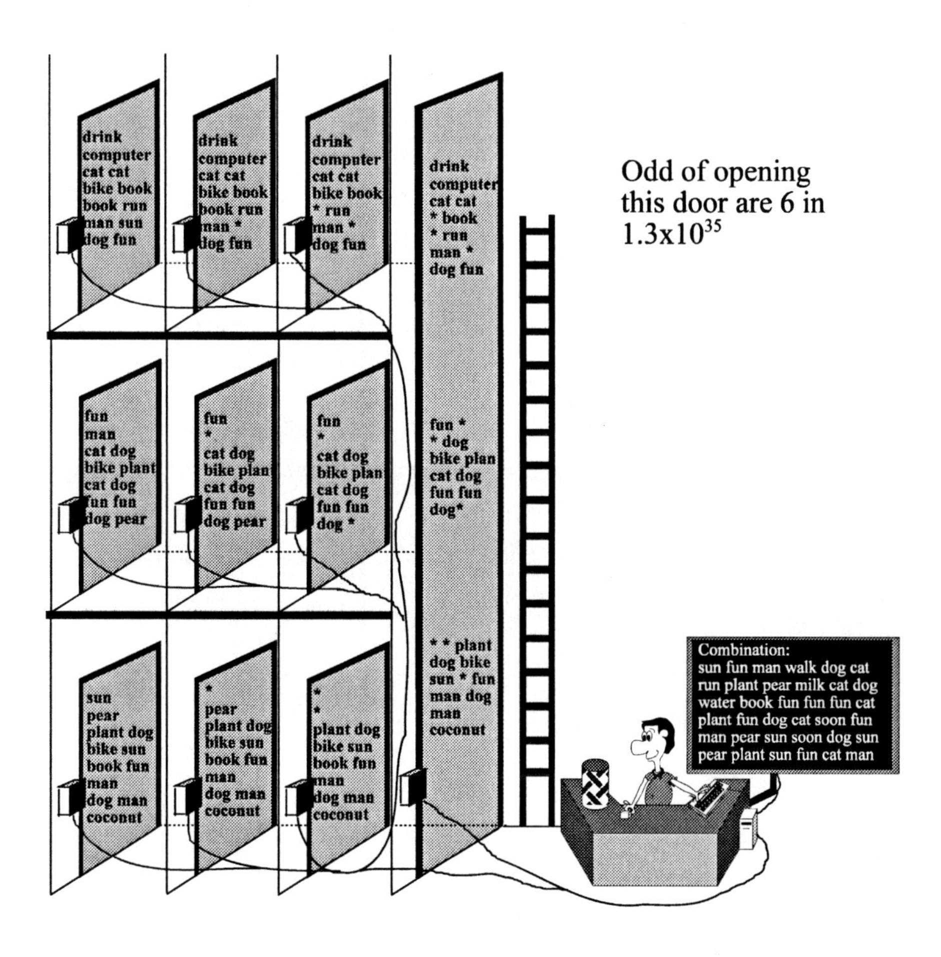

The odds for opening the first door are found as follows: this door requires 27 letters to be correct, and each letter has a 1 in 20 chance of arising by chance. So the odds of finding the combinations are 1 in 20^{27} tries. The odds are actually slightly better. The 27 required letters on this first door can be divided into 3 groups, and if the group associated with the bottom floor is switched with the one on the top, then the door should still open because all 3 infons still exist.

These three groups can be arranged into 6 different combinations because each series of doors can be moved to another story. For example, the doors on the first story can be moved to the second, and those on the second are moved to the first. Thus, a total of 6 outcomes will open the first door with 20^{27} attempts. This corresponds to 6 times in 1.3×10^{35} tries or a 1 in 2.2×10^{34} chance.

This example can be extended to cover more complex systems by imagining a scientist in a taller building. Perhaps, the first door is 10 stories high instead of three. In any case, the odds of the scientist opening the first door are vanishingly small.

Examples of irreducible complexity exist everywhere in life. The best examples are the metabolic pathways that synthesize complex chemicals. Metabolic pathways are a series of chemical reactions regulated by enzymes. These chemical reactions may create biological molecules like adenine, cytosine, uracil, thymine, glucose or ribose. The next chapter will consider one such pathway. It is the pathway that all living things with the exception of few parasitic organisms use to make adenine.

Reference:

Behe, Darwin's Black Box: the Biochemical challenge to Evolution, Touchstone, 1996.

Chapter 13: Nucleic Acid Synthesis: Adenine

Science has yet to find a solution for chemical evolution or the origin of life, and many scientists realize that this is the case. It seems that the puzzle just gets harder as the years pass. In the 1950s, most of the scientists investigating the origin of life probably thought that they would solve the problem. Today, it is clear that this is not going to happen. The self replicating RNA molecule theory persists only because nobody has proposed a better hypothesis. Science is trapped because by definition, science is only allowed to consider naturalistic explanations. Science may never find the solution.

While most scientists realize that the origin of life is a rather formidable barrier, few realize that the barrier associated with the creation of the first genes and proteins is just as high. The next two chapters will explore this topic.

Nucleic Acids are Critical to Life

The first living organism either needed to find adenine and ATP in its surroundings or synthesize these two chemicals from more abundant chemicals found in the soup. If the first living molecule used impA instead of ATP, then it would have to either synthesize impA or find impA. A similar argument can be made for the other activated nucelotides required by life to replicate (GTP, UTP, CTP, TTP, impG, impU or impC).

In the laboratory, chemists can create adenine in an enclosed vessel from a concentrated solution of hydrogen cyanide and ammonia. The first form of life would be destroyed by these conditions, and because there is simply no way to concentrate volatile gases in an open environment (like the primitive earth), the conditions required to create large yields of adenine in the lab do not model the conditions on the primitive earth. So finding enough adenine in the primordial soup to support life seems very unlikely.

Miller has suggested that adenine was synthesized from hydrogen cyanide at very cold temperatures as water freezes, melts and re-freezes. While this process may have produced some adenine, if the entire body of water ever melts, then the adenine will again be too dilute. Furthermore, the freezing cycles described above are limited to small ponds because the salts present in the ocean interfere with adenine synthesis. Despite these concerns, adenine has been found in very small concentrations in some meteorites. So nature can make it in small quantities; as a result, some adenine was probably dissolved in the primitive ocean.

The problem is one of dilution. If the first living molecule does not have the ability to synthesize adenine and ATP, then it will have to search the ocean for decades, collecting these chemicals before it can replicate. Since the first living molecule must be simple, it would be more reasonable to speculate that the necessary chemicals float by the living molecule, and it simply collects them. Such a molecule may have to wait several thousand years to accumulate the ATP that it needs to replicate just once. This is not a reasonable model for a living system. The lifetime of the chemicals that make up the living molecule (assuming it is an RNA molecule) is very short (see chapter 9). So such a molecule will not exist long enough to replicate once.

Suppose instead that the primordial soup exists locally in a small pond or puddle. In theory, with the appropriate concentration mechanisms in place, such a puddle may have a concentration of nucleic acids many orders of magnitude higher than that of the primitive ocean. If life evolves in the puddle, it will quickly deplete the supply of free nucleic acids as it replicates. It will then run into the dilution problem outlined in the previous paragraph.

This argument and several others like it suggest that life was never as simple as many scientists have theorized. The enzymes that synthesize adenine and the other bases (guanine, uracil, cytosine, and thymine) almost certainly had to exist either before or coincidentally with the origin life. The same can be said for the enzymes that synthesize ATP and the other activated nucleotides. The difficulties with chemical evolution demand that this be the case.

Furthermore, there is quite a bit of evidence that suggests that these enzymes did exist. Every single living thing (with the exception of a few parasites who have lost the genes) shares the same genes that encode the enzymes responsible for making adenine. This means that the genes responsible for the synthesis of adenine and the other bases existed in the common ancestor to all living things, 2.5 to 4 billion years ago.

Figure 13.1: The Tree of Life

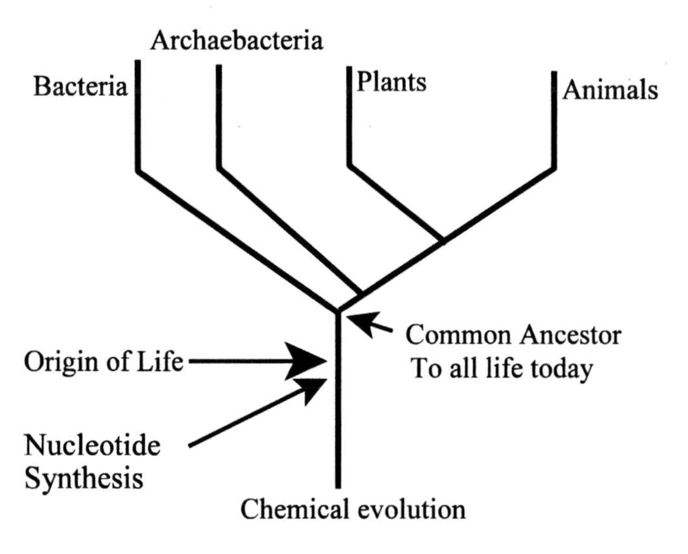

Because life on the primitive earth was not possible without the genes responsible for making nucleic acids, these genes may have actually proceeded life (figure 13.1). Even if life originated first, the problem remains unsolvable. The metabolic pathway that will be discussed next is perhaps the best example of an irreducibly complex system that can be found in life. The synthesis of adenine requires 11 enzymes. If a single enzyme is missing, the yield of adenine is zero. Therefore, all 11 enzymes must evolve together and become at least marginally functional before natural selection can preserve and optimize the system. The implication is that the very first step required to create the molecular knowledge is insurmountable. Chance and natural selection cannot explain the evolution of this metabolic pathway.

Adenine Synthesis in Life

Life synthesizes adenine directly on the ribose ring. Eleven enzymes are required to make this happen. If a single enzyme is not functional, then no adenine can be synthesized. It is an irreducibly complex system. Before natural selection can help, all eleven enzymes must be marginally functional. The enzyme names and number of amino acids per enzyme are listed in table 13.1. The enzyme names are not important for this discussion and are given only for reference.

Table 13.1 Enzymes Required for Adenine Syntheses

Enzyme Name	Amino Acids
amidophosphoribosyly transferase	474
phosphoribosylyglycinamide synthase	416
phosphoribosylyglycinamide formyltransferase	195
phosphoribosylyformylglycinamide synthase	1295
phosphoribosylaminoimidazole synthase	346
phosphoribosylaminoimidazole carboxylase	162
phosphoribosylaminoimidazole - succinocarboxamide synthase	239
adenylosuccinate lyase	456
phosphoribosylaminoimidazolecarboxamide & formyltransferase and IMP cyclohydrolase (2 enzymes in one protein)	512
adenylosuccinate synthase	432

Total amino acids = 4,527

Preliminary Calculation of the Information and Knowledge

The analysis undertaken in chapters 4 and 5 could be repeated for all of these enzymes, but this would take several weeks to analyze the data and then another week to format the data in a presentable fashion.

Is there a simpler way? In chapter 5, the analysis revealed that the B chain of insulin contains 280.5 bits of information and 211 bits of knowledge. Since this chain contains 30 amino acids, the average information per amino acid is 280.5/30 = 9.35 bits per amino acid and 211/30 = 7 bits of knowledge per amino acid.

To find an answer quickly without any long drawn out mathematical analysis, the molecular knowledge required for life to synthesize adenine is given by: 7 bits per amino acid x 4527 amino acids = 31,689 bits. Refer to table 13.1 for the total amino acids.

Notice that the above calculation uses molecular knowledge instead of information. Nevertheless, the above calculation assumes that the average knowledge of a small highly conserved protein like insulin is representative of large enzymes. This assumption is incorrect, and so the number of bits calculated above is too large.

In a large enzyme, patches of amino acids that are located near each other on the three-dimensional structure of the protein tend to either be highly conserved or highly variable. The highly conserved patches will likely show a conservation pattern similar to insulin. Thus, the average knowledge per amino acid (based on the analysis of insulin) can only be safely applied to these highly conserved patches.

3-D Structure of Several Key Enzymes

The protein data bank houses a database which contains the 3-D structures of many proteins. All of the proteins involved in adenine synthesis can be found in this online database.

Another program called CONSURF will accept a protein database entry as input and generate a script that can be used to color the amino acids by their variability. The default of this program is to color highly conserved amino acids as shades of purple and weakly conserved amino acids as blue and white. The script so generated was modified so that all of the purple atoms appear as a shade of gray. Blue and off white were also modified to appear as white. The modifications were necessary so that the images would print correctly in black and white. The modified script was used to color several of the enzymes used in the synthesis of adenine (figures 13.2 - 13-5).

The black and gray regions in the following figures are regions of high conservation. These regions contain most of the information. Between twenty to thirty percent of the amino acids found in these proteins belong to a highly conserved region. In these regions, the knowledge per amino acid should be similar to that of insulin.

In chapter 5, because of the presence of non-biological amino acids in the primordial soup, 2 bits of primordial information are assigned to every amino acid that is used by life. So if these proteins evolved in the primordial soup, the white regions in the following figures must contribute at least 2 bits of information, and in this special case, because primordial information is a form of knowledge, it must be added to molecular knowledge to calculate the total knowledge, and this total can be related to a probability (see chapter 5).

If on the other hand, these proteins evolved after the genetic code was in place, then the white regions would not contribute much to knowledge. This scenario seems unlikely because life needs adenine to replicate, and chemical evolution does not create large quantities of adenine.

Figure 13.2: Amidophosphoribosyl Transferase (474 amino acids)

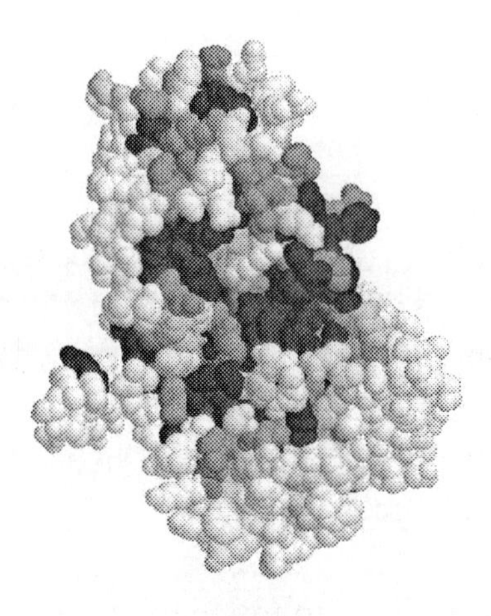

Figure 13.3: Phosphoribosylaminoimidazole carboxylase (162 amino acids)

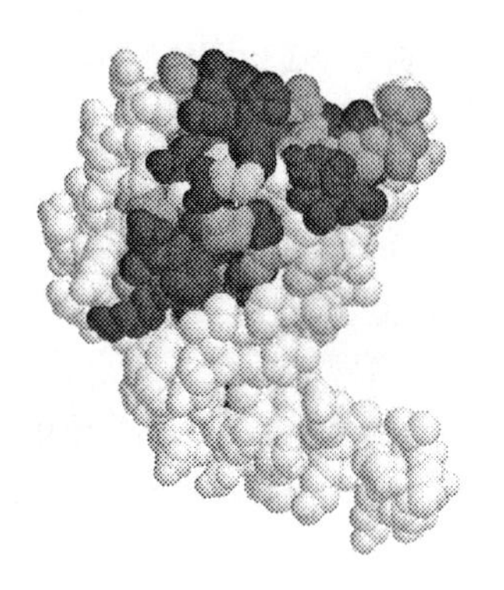

Figure 13.4: Adenylosuccinate Lyase (456 amino acids)

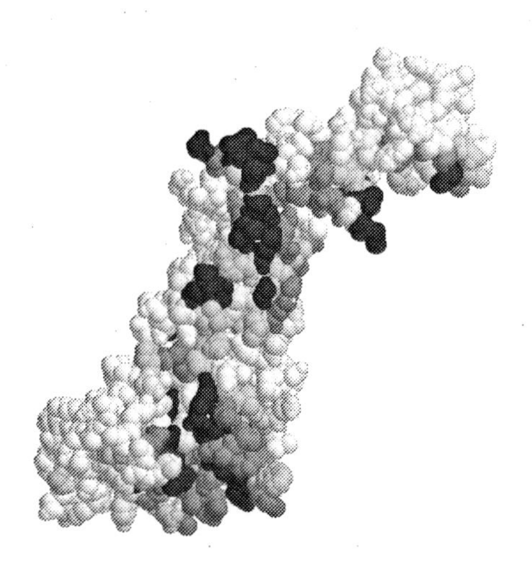

Figure 13.5: Phosphoribosylaminoimidazolecarboamide and IMP Cyclohydrolase (2 enzymes in one - 512 amino acids)

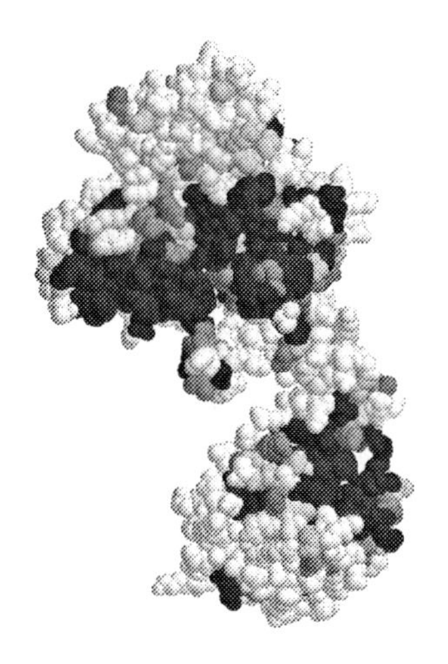

Calculation of the Knowledge and Information

The total number of amino acids found in these enzymes is not nearly as important as the number of higher conserved amino acids. From figures 13.2 to 13.5, it appears that approximated 20-30% of the amino acids in these proteins belong to highly conserved patches.

So of the 4527 amino acids found in these enzymes, assume that only 20% contribute to molecular knowledge. With this assumption, the numbers found in the preliminary calculation now need to be multiplied by 20%.

Molecular Knowledge = 4527 x .2 x 7 bits per amino acid= 6,338 bits (black and gray regions)

Primordial
Information = 4527 x .8 x 2 bits per amino acid = 7,243 bits
(white regions)

Total number of bits: 13,581 bits.
Odds for success: 1 in 2^{13581} of 1 in 2 x 10^{4088}

In figure 13.6, the total number of bits calculated above is shown as the first step. Notice that primordial information and knowledge are additive. This must be true because primordial information is a form of knowledge - the knowledge to use only the twenty biological amino acids in the soup (see chapter 5).

The above calculation assumes that the 11 enzymes responsible for synthesizing adenine evolved before or coincidently with the origin of life. Consider the case where the opposite is true. Suppose that the first living molecule obtained adenine from the soup. Could these 11 enzymes evolve after the genetic code is in already in place? The number of bits is now as follows: knowledge = 4527 amino acids * 20% * 2.5 bits per amino acid = 2,263.5 bits. Refer to page 102. Notice that after the genetic code exists primordial information equals zero, and molecular knowledge equals total knowledge.

Figure 13.6: The Evolution of Molecular Knowledge Required to Synthesize Adenine in the Soup and with the Genetic Code

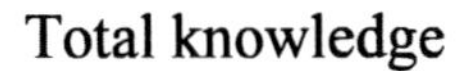

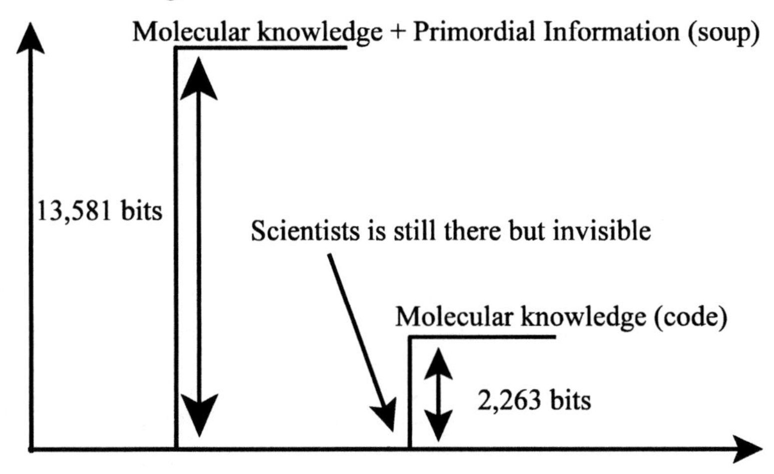

At this scale, the scientist is so small that he is not visible in the picture. There is no way he can climb either wall.

Adenine synthesis is perhaps the best example of an irreducibly complex system that can be found in life, but there are many other metabolic pathways that need all of their enzymes to at least be partially functional before they can have any function at all.

Many proteins are complexes of smaller peptides, and many of these proteins are themselves irreducibly complex. This will be the subject of the next chapter.

Did Life Arise All at Once?

Researchers have attempted to reconstruct the tree of life by comparing the RNA sequences found in ribosomes (ribosomal RNA) in many diverse species of bacteria, archaebacteria, fungi, protozoans, animals and plants. Others have done the same by comparing the amino acid sequences of ancient proteins that are found in all living things. The results have been ambiguous.

While few researchers have been bold enough to suggest that life arose all at once, the evidence from the reconstruction experiments suggest that it did. The tree as it is conventionally drawn in figure 13.1 suggests that bacteria diverged first, and then archaebacteria second. This means that archaebacteria are more closely related to man than bacteria. The drawing in figure 13.1 is based on the analysis of ribosomal RNA, but this is not consistent with the protein comparisons that show that archaebacteria and bacteria share many of the same genes that are not found in higher forms of life.[1]

"If these two prokaryotic groups span the primary phylogenic divide and their genes are vertically (genealogically) inherited, then the universal ancestor must have had all of these genes, these many functions: This distribution of genes would make the ancestor a prototroph with a complete tricarboxylic acid cycle, polysaccharide metabolism, both sulfur oxidation and reduction, and nitrogen fixation; it was motile by means of flagella; it had a regulated cell cycle and more. This is not the simple ancestor, limited in metabolic capabilities, that biologist originally intuited. That ancestor can explain neither the the broad distribution of diverse metabolic functions nor the origin of early autotrophy implied by this distribution. The ancestor that this broad spectrum of metabolic genes demands is totipotent, a genetically rich and complex entity, as rich and complex as any modern cell - seemingly more so." - Carl Woese, The Universal Ancestor.

"For instance, transcription, translation and splicing machineries of the archaebacteria resemble those of the eukaryotes, while the majority of the functional genes, coding primarily for metabolic enzymes, transport systems and enzymes of cell wall biogenesis, resemble the eubacterial ones. Microbiologists have reviewed a number of possible explanations for this mosaic, but none of them seems to be, at the present time, particularly convincing." - Mayr, "Two empires of Three,"PNAS, 1998.

The ribosomal RNA data does not agree with the protein data. The ribosomal data suggest that the archaebacteria are more closely related to man than bacteria. Yet the proteins tell another story. Woese resolved the dilemma by suggesting that the common ancestor was not a single cell, but that instead a collection of many different very simple cells. None of these simple cells had the ability to live alone, but together by sharing and transferring genes, they managed.

This is an interesting idea, but it does not change the probability. One of the nice features about information is that it is additive. So one complex protein containing 500 bits of knowledge has the same chance of evolving as 10 smaller proteins containing 50 bits each. So the probability for evolution does not depend on whether life emerged as one cell or a colony of cells. While the first cells may have shared information more freely than modern cells, this observation does not solve any of the problems. The only reason to even propose a colony of simple cells is that it allows each of the cells to be simpler than cells today. Nevertheless, this reasoning is flawed because the colony as a whole is not simpler. For example, eleven enzymes are required for adenine synthesis, and all of these must exist before life can make adenine. If nine cells possess one of the required enzymes, then the colony will not be able to make any adenine because adenine synthesis requires 11 enzymes.

Mayr perhaps offered a better explanation by suggesting that life is composed of only two kingdoms not three. This of course still implies that the common ancestor to all life was not simple, but possessed a multitude of genes. It very well may have been more complex than any living cell found today.

The difficulties associated with chemical evolution almost demand that the first living thing be robust and complex. No organism whose genetic structure is based on RNA or DNA can replicate itself unless it can synthesize the bases required to do so (adenine, cytosine, guanine, thymine and uracil), and if ribose was used as the backbone in the first DNA/RNA, then it too must be synthesized. This implies photosynthesis and the Calvin cycle must be present. Furthermore, activation of nucleotides to create molecules like ATP requires that this first living molecule tap a plentiful energy source. The simplest living organisms today are parasitic. They obtain nourishment by absorbing nutrients from their host, and this process allows them the luxury of not having many genes that would otherwise be necessary. Given the dilute concentration of biological precursors in the soup (if it existed), the first living thing would not be able to rely on the soup for nourishment. The first form of life was a complete living cell with many if not all of the capabilities found in life today.

References:

1) Woese, "Interpreting the Universal Phylogenic Tree," PNAS, 97: 8392-9396, 2000.

2) Woese, "The Universal Ancestor," PNAS, 95:6854-68-59, 1998.

3) Mayr, "Two Empires or Three," PNAS, 95:9720-9723, 1998.

4) Meyer, "The Origin of Bilogical Information and the Higher Taxonomic Categories," Proc. Of the Biological scociety of Washington, 117:213-239, 2004.

5) Glaser F./ Pupko T., Paz I., Bell R.E., Becher D., Martz E., Ben-Tal N., Consurf: Identification of Functional Regions in Proteins by Surface Mapping of Phylogentic Information, Bioinformatics, Vol. 19, no, 1, 2003 pp163-164.

Chapter 14: ATP Synthesis

The last chapter dealt with the synthesis of adenine, but it was not a complete analysis because even if the enzymes required to create adenine existed, they would have no way to power the chemical reactions that they facilitate. These enzymes would be like a gas powered car with no gas, or a solar powered car with no sun and no battery.

The last chapter presented the argument that the concentration of adenine in the primitive ocean was very dilute; therefore, any proposed form of life that uses RNA and DNA to replicate must be able to synthesize adenine. The same argument is even more compelling with ATP. The concentration of ATP cannot exceed that of adenine because adenine is needed to make ATP. Unlike adenine, ATP contains high energy bonds. This means that ATP would have a very short lifetime. ATP would undoubtably decay to AMP in a matter of days. So the concentration of ATP in the primitive ocean or in a localized soup would have been negligible.

Figure 14.1: ATP has Two High Energy Bonds

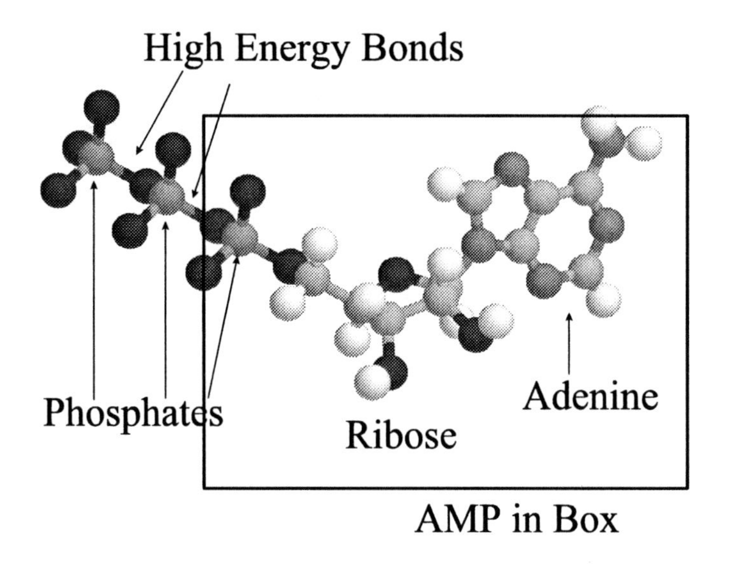

ATP stands for adenine triphosphate. Figure 14.1 shows that ATP is composed of one adenine molecule, one ribose molecule, and 3 phosphate groups. The high energy bonds are located between the phosphate groups.

All living things use ATP for energy. This means that the capability to synthesize ATP arose before the common ancestor diverged to create all the branches in the tree of life. Both adenine and ATP synthesis were required before life could emerge. Life that has to wait several hundred thousand years to replicate is not a reasonable model for the first living cell. Not only is such an organism replicating too slowly to evolve in the time allotted, but it certainly would not replicate at all as it would be destroyed before it acquired the necessary chemicals. The first living organism must have been able to synthesize both adenine and ATP.

Figure 14.2: The Tree of Life

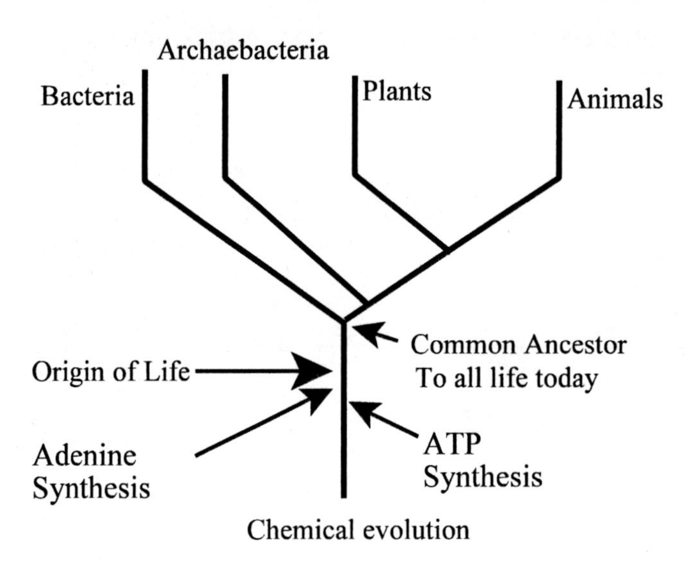

In figure 14.2, ATP synthesis is shown coincidently with adenine synthesis. On purely theoretical grounds, the two processes must have evolved at the same time. It is one of the many chicken or the egg paradoxes that plague origin of life theories.

The synthesis of a chemical like adenine from the chemicals readily available of the primitive earth would result in a decrease in entropy. So this cannot happen unless an energy source is used to drive the chemical reaction. Plenty of energy sources were available on the primitive earth, but unless some system is in place that can use these energy sources to drive a chemical reaction, the energy sources are of no value. Life knows how to use energy sources to drive chemical reactions. That is life can take the energy stored in a chemical like ATP, and use this energy to synthesize another chemical, like adenine (the enzymes that create adenine actually require 5 ATP molecules to drive the process). How can life possibly make ATP if the concentration of adenine is so low that the first living cell only comes into contact with an adenine molecule every hundred years or so? The model is only plausible if both capabilities emerge at the same time.

The implication is that the knowledge calculated in the last chapter is too small. For the system to function, the enzymes in the last chapter also need ATP. Thus, the total system must include the enzymes required to make adenine, as well as the additional enzymes required to make ATP.

Life creates ATP by a process called oxidation. Oxidation will be the subject of the next section.

Oxidation Releases Energy

Oxidation involves electron transfer. Electrons are transferred from chemicals that do not want them (electron donors) to chemicals that desire them (electron acceptors). In figure 14.3, the process of oxidation is illustrated with a mechanical example. An electron donor (the hand) places the electron (a very heavy ball) on the ramp. The electron rolls down the ramp and activates the lever which lifts the weight that represents ATP in figure 14.4. In this way, the potential energy of the electron is used to do work. In the mechanical example, it lifts the weight. In life, the energy released by oxidation is used to create ATP.

Figure 14.3: A Mechanical Example of Oxidation

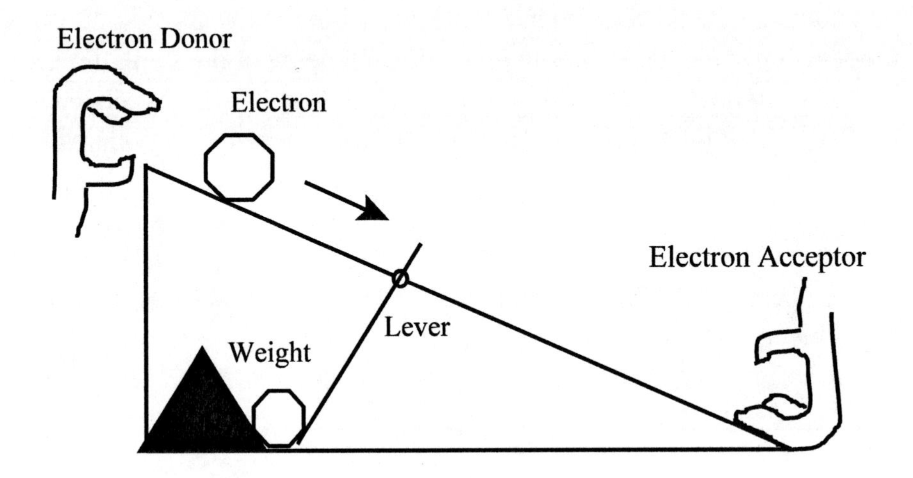

Figure 14.4: The Electron Does Work When it Lifts the Weight

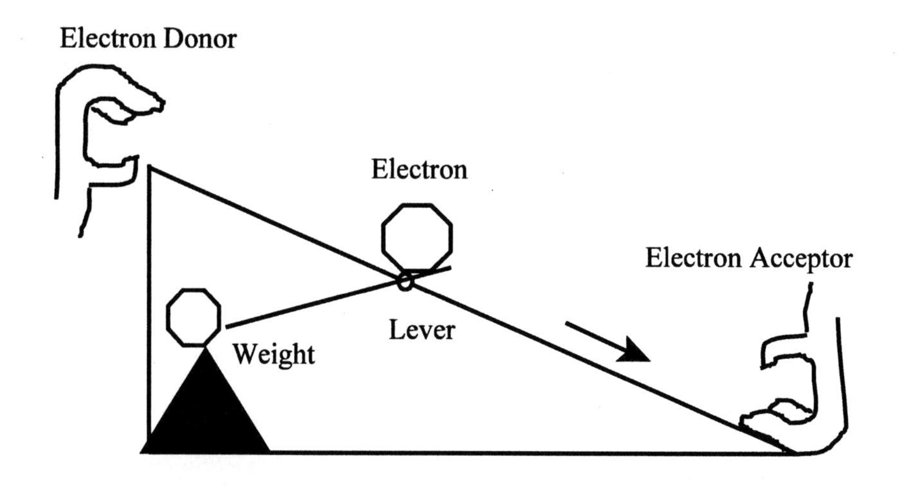

Figure14.5: Final State - The Weight Can Now Perform Work

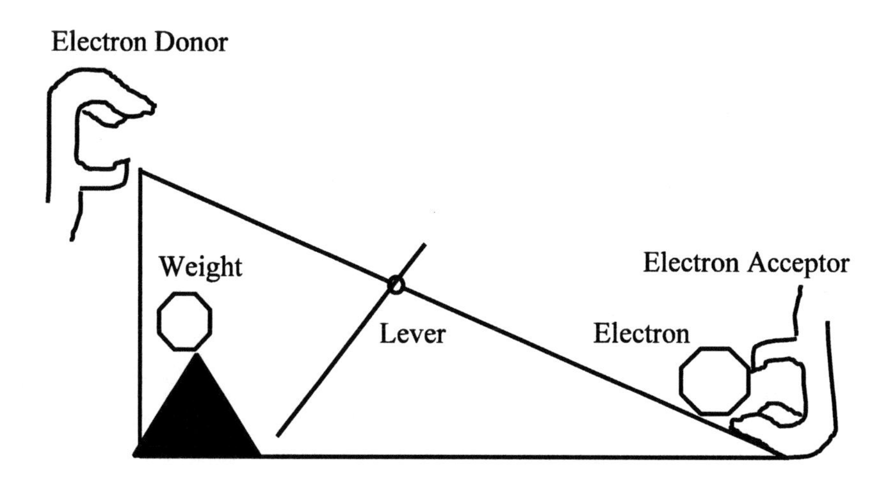

In figure 14.5, the weight is left on top of a smaller ramp. It too can now perform work. In life, the work is often maximized by using a series of several electron donors and acceptors. Bacteria in particular are very good at using many different chemicals for both acceptor and donor. This allows them to consume all kinds of chemicals for food. In higher life, the primary electron donor is glucose (sugar) and the final acceptor is oxygen.

ATP Created by Proton Gradients

Most ATP in both bacteria and in higher life is generated by a proton gradient. An electron donor, donates an electron to a series of carriers imbedded in a membrane. As electrons are transferred from donors to acceptors, the energy release is used to pump protons across the membrane, figure 14.6. Another very complex enzyme called ATP synthase then uses this proton gradient to create ATP from ADP. Because it is found in all living things, ATP synthase must have evolved very early in the evolutionary process.

Figure 14.6 Electron Transfer Drives a Proton Pump

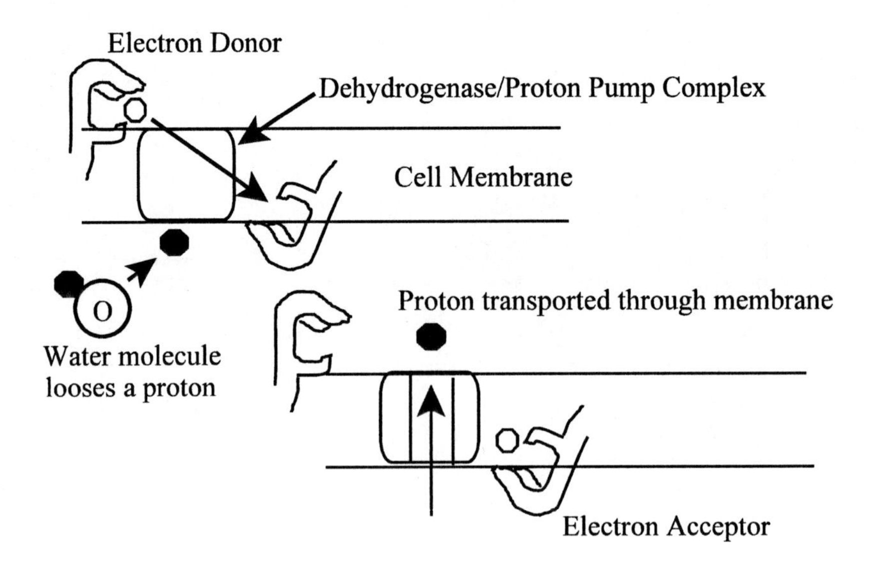

Many authors have suggested that one of the various bacteria that consume hydrogen, methane, iron or sulfur was probably one of the first forms of life, but these metabolic pathways are very complex and contain too much knowledge to be the first route to synthesize ATP. They all use ATP synthase in combination with the proton pump shown in figure 14.6 to generate ATP. Furthermore, ATP synthase is one of the most complex proteins found in life. There must be a simpler solution.

A class of enzymes called dehydrogenases facilitate oxidation and in many cases, capture the energy released. In figure 14.6, the dehydrogenase powers the proton pump, but many dehydrogenase enzymes are capable of directly creating high energy phosphate bonds. Today, most metabolic pathways create ATP using two or more steps. In step 1, the dehydrogenase captures the energy released by oxidation and uses this energy to add a high energy phosphate bond to an intermediate chemical. Another enzyme called a kinase then transfers this high energy phosphate from the intermediate to ADP creating ATP. This is generally a two-step process because it is more efficient at capturing the energy released by oxidation. In theory at least, there is no reason that the dehydrogenase could not create ATP directly. This path seems much more promising for origin of life theories because it only requires a single enzyme. The next section will explore this idea.

ATP Synthesis with a Single Enzyme

This section will consider a dehydrogenase that is involved in glycolysis (the process life uses to metabolize sugar), Glyceraldehyde 3-P dehydrogenase or G3PD for short. G3PD facilitates an oxidation reaction and uses the energy released to create a high energy phosphate bond. This high energy phosphate is subsequently transferred to ATP by another enzyme. In theory, an enzyme with a molecular knowledge on par with G3PD should be able to create ATP from ADP. So this section will calculate the knowledge in G3PD, and use it as a benchmark to model a hypothetical enzyme that may have existed on the primitive earth.

G3PD Knowledge

The same technique used in chapter 13 could be used to calculate the knowledge in G3PD, but since this is a single protein the techniques discussed in chapters 4 and 5 will be used. The advantage of this approach is that it requires fewer assumptions.

The knowledge will first be calculated assuming G3PD arose in the primordial soup. The analysis will then be repeated assuming that the protein evolved after the genetic code was in place. Because G3PD is a typical enzyme of about average size, there is every reason to believe that it is representative of most other enzymes.

Fourteen amino acid sequences for G3PD were downloaded from this web site: http://us.expasy.org/sprot/. Seven are from higher animals or plants, and six are from bacteria. Every effort was made to pick the sequences to maximize the diversity. The sequences were aligned using the computer program, Clustal X. To obtain Clustal X visit: http://www-igbmc.u-strasbg.fr/BioInfo/. Once aligned the columns were scored by amino acid identity, and this file was saved as an output of the program. The file was imported into a spreadsheet, filtered, and manipulated. In the end, a spreadsheet was obtained that calculates knowledge. The techniques to do this manually are described in chapters 4 and 5. An automated system is necessary here because G3PD contains many more amino acids that insulin.

The multiple sequence alignment is shown on the next page. The dashes are gaps inserted by Clustal to align the proteins.

CLUSTAL X (1.8) MULTIPLE SEQUENCE ALIGNMENT

File: C:clustalgyceraldehyde_dehydrogenase.ps **Date: Fri Apr 15 20:55:40 2005**
Page 1 of 1

```
                  :.::*** ***    *              .: .:** .      :  :  : *:                    :  :                    .* .: :    .        *.:* *          *   .  .:::::...         . ..
         human --GKVKVGVDGFGRIGRLVTRAAFNSGK---VDIVAINDPFIDLHYMVYMFQYDSTHGKFHG---TVKAEDGK--LVIDG---KAITIFQERDPENIKWGDAGTAYVVESTGVFTTMEKAGAHLKGGAKRIVISAPSADA---PMFVMGV   134
       Chicken ----VKVGVNGFGRIGRLVTRAAVLSGK---VQVVAINDPFIDLNYMVYMFKYDSTHGHFKG---TVKAENGK--LVING---HAITIFQERDPSNIKWADAGAEYVVESTGVFTTMEKAGAHLKGGAKRVIISAPSADA---PMFVMGV   132
          frog ----VKVGINGFGCIGRLVTRAAFDSGK---VQVVAINDPFIDLDYMVYMFKYDSTHGRFKG---TVKAENGK--LIIND---QVITVFQERDPSSIKWGDAGAVYVVESTGVFTTTEKASLHLKGGAKRVVISAPSADA---PMFVVGV   132
         trout -MSDLCVGINGFGRIGRLVLRACLQKG----IKVVAINDPFIDLQYMVYMFKYDSTHGRYKG---EVSMEDGK--LIVDD---HSISVFQCMKPHEIPWGKAGADYVVESTGVFLSIDKASSHIQGGAKRVVVSAPSPDA---PMFVMGV   134
       lobster ----SKIGINGFGRIGRLVLRAALSCG----AQVVAVNDPFIALEYMVYMFKYDSTHGVFKG---EVKMEDGA--LVVDG---KKITVFNEMKPENIPWSKAGAEYIVESTGVFTTIEKASAHFKGGAKKVVISAPSADA---PMFVCGV   131
           Fly ---MSKIGINGFGRIGRLVLRAAIDKG----ANVVAVNDPFIDVNYMVYLFKFDSTHGRFKG---TVAAEGGF--LVVNG---QKITVFSERDPANINWASAGAEYIVESTGVFTTIDKASTHLKGGAKKVIISAPSADA---PMFVCGV   132
          corn -MGKIKIGINGFGRIGRLVARVALQSED---VELVAVNDPFITTDYMTYMFKYDTVHGHWKHSDITLKDSKTL--LFGD----KPVTVFGIRNPEEIPWGEAGAEYVVESTGVFTDKDKAAAHLKGGAKKVVISAPSKDA---PMFVVGV   137
Dinoflagellate --MPPTMGINGFGRIGRLVFRAALANPD---IEVKAVNDPFMDVKYMVYQLKYDSVHKRFPG-TISTKEVDGKEFLVVDG---KDVQVFHEKDPASIPWGAAGANYICESTGVFTAQEKAELHIKGDAKKVIISAPPKDAV--PIYVVGV   139
        E_coli ---TIKVGINGFGRIGRIVFRAAQKRSD---IEIVAIND-LLDADYMAYMLKYDSTHGRFDG---TVEVKDGH--LIVNG---KKIRVTAERDPANLKWDEVGVDVVAEATGLFLTDETARKHITAGAKKVVMTGPSKDNT--PMFVKGA   133
    Haloarcula MSKPVRVGLNGFGRIGRNVFRASLHNDD---VEIVGIND-VMDDSEIDYFAQYDTVMGELEG----ASVDDG--VLTVEG-TDFEAGIFHETDPTQLPWEDLDVDVAFEATGIFRTKEDASQHLDAGADKVLISAPPKGDEPVKQLVYGV   139
   Helicobacter ---MINIAINGFGRIGRSIMRVALQHKYKEHISIVAIND-INDWEILSYLLENDTTHRTLPF---EVSHTSNTLILKNNHNTYPPIRTFNHSNPKELDFAACGADIVIESSGQFLDSKQLTHHCEKGIKRVILSATPTDN--MPTFALGV   141
       Aquifex --MAIKVGINGFGRIGRSFFRASWGREE---IEIVAIND-LTDAKHLAHLLKYDSVHGIFKG---SVEAKDDS--IVVDG---KEIKVFAQKDPSQIPWGDLGVDVVIEATGVFRDRENASKHLQGGAKKVIITAPAKNPD--ITVVLGV   134
   Clostridium ---MTKVAINGFGRIGRLALRRILEVPG---LEVVAIND-LTDAKMLAHLFKYDSSQGRFNG---EIEVKEGA--FVVNG---KEVKVFAEADPEKLPWGELGIDVVLECTGFFTKKEKAEAHVRAGAKKVVISAPAGNDL--KTIVFNV   133
      Bacillus ---AVKVGINGFGRIGRNVFRAALNNPE---VEVVAVND-LTDANMLAHLLQYDSVHGKLDA---EVSVDGNN--LVVNG---KTIEVSAERDPAKLSWGKQGVEIVVESTGFFTKRADAAKHLEAGAKKVIISAPANEED--ITIVMGV   133
         ruler 1.......10........20........30........40........50........60........70........80........90.......100.......110.......120.......130.......140.......150

               *              *.****:*.::*  *::.: *  :   .  :   :*: *   *   :*             *     *       .::*  ***:::    ::* :.**: * : ***       : ::    :        .  ::     .   *:
         human NHFKYAN-SLKIISNASCTTNCLAPLAKVIHDHFGIVEGLMTTVHAITATQKTVDSPSG-KLWRGGRGAA---QNLIPASTGAAKAVGKVIPELDGKLTGMAFRVPTANVSVLDLTCRLEKP-AKYDDIKKVVKEASEGP-------LKG   271
       Chicken NHEKYDK-SLKIVSNASCTTNCLAPLAKVIHDNFGIVEGLMTTVHAITATQKTVDGPSG-KLWRDGRGAA---QNIIPASTGAAKAVGKVIPELNGKLTGMAFRVPTPNVSVVDLTCRLEKP-AKYDDIKRVVKAAADGP-------LKG   269
          frog NHEKYEN-SLKVVSNASCTTNCLAPLAKVINDNFGIVEGLMTTVHAFTATQKTVDGPSG-KLWRDGRGAG---QNIIPASTGAAKAVGKVIPELNGKITGMAFRVPTPNVSVVDLTCRLQKP-AKYDDIKAAIKTASEGP-------MKG   269
         trout NEDKFDPSSMTIVSNASCTTNCLAPLAKVIHDNFGIEEALMTTVHAYTATQKTVDGPSA-KAWRDGRGAH---QNIIPASTGAAKAVGKVIPELNGKLTGMAFRVPVADVSVVELTCRLSRP-GSYAEIKGAVKKAAEGP-------MKG   272
       lobster NLEKYSK-DMTVVSNASCTTNCLAPVAKVLHENFEIVEGLMTTVHAVTATQKTVDGPSA-KDWRGGRGAA---QNIIPSSTGAAKAVGKVIPELDGKLTGMAFRVPTPDVSVVDLTVRLGKE-CSYDDIKAAMKTASEGP-------LQG   268
           Fly NLDAYKP-DMKVVSNASCTTNCLAPLAKVINDNFEIVEGLMTTVHATTATQKTVDGPSG-KLWRDGRGAA---QNIIPASTGAAKAVGKVIPALNGKLTGMAFRVPTPNVSVVDLTVRLGKG-ASYDEIKAKVQEAANGP-------LKG   269
          corn NEDKYTS-DVNIVSNASCTTNCLAPLAKVIHDNFGIVEGLMTTVHAITATQKTVDGPSA-KDWRGGRAAS---FNIIPSSTGAAKAVGKVLPDLNGKLTGMSFRVPTVDVSVVDLTVRIEKG-ASYEDIKKAIKAASEGP-------LKG   274
Dinoflagellate NHEDYKT-DRHSWSNASCTTNCLAPLTKVVHEKFGLLEGLMTTVHAMTATQLTVDGPS--VAARTGVVAAAHPQNIIPSSTGAAKAVGKVIPAVNGKLTGMAFRVPTPDVSVVDLTCRIEKA-AKYDDIVAAIKEAAAGP-------MSG   278
        E_coli NFDKYAG--QDIVSNASCTTNCLAPLAKVINDNFGIIEGLMTTVHATTATQKTVDGPSH-KDWRGGRGAS---QNIIPSSTGAAKAVGKVLPELNGKLTGMAFRVPTPNVSVVDLTVRLEKA-ATYEQIKAAVKAAAEGE-------MKG   269
    Haloarcula NHDEYEG--EDVVSNASCTTNSITPVAKVLDEEFGINAGQLTTVHAYTGSQNLMDGPNG--KPRRRRAAA---ENIIPTSTGAAQAATEVLPELEGKLDGMAIRVPVPNGSITEFVVDLDDD-VTESDVNAAFEEAAAG-------DLEG   274
   Helicobacter NHTLYAK--ESIISNASCTANALAPLCKIIDECFGIEVASFCIMHSYTNEQSLLDSVYP-TNKRRSRAAA---QNIIPTHTGATRTLMRILPELEGKIDGHSVRVPISNVLLIDITFNLSRQ-TTITELNNALINASKTS-------MKG   277
       Aquifex NEEKYNPKEHNIISNASCTTNCLAPCVKVLNEAFGVEKGYMVTVHAYTNDQRLLDLPHK--DFRRARAAA---INIVPTTTGAAKAIGEVIPELKGKLDGTARRVPVPDGSLIDLTVVVNKAPSSVEEVNEKFREAAQKYRESGKVYLKE   279
   Clostridium NNEDLDGTET-VISGASCTTNCLAPMAKVLNDKFGIEKGFMTTIHAYTNDQNTLDGPHRKGDFRRARAAA---VSIIPNSTGAAKAIAQVIPELKGKLDGNAQRVPVPTGSVTELISVLKKN-VTVEEINAAMKEAAN----------E   267
      Bacillus NEDKYDAANHDVISNASCTTNCLAPFAKVLNDKFGIKRGMMTTVHSYTNDQQILDLPHK--DYRRARAAA---ENIIPTSTGAAKAVSLVLPELKGKLNGGAMRVPTPNVSLVDLVAELNQE-VTAEEVNAALKEAAEG-------DLKG   270
         ruler .......160.......170.......180.......190.......200.......210.......220.......230.......240.......250.......260.......270.......280.......290.......300

                .    .     ** *   .          .          ..        :.:   :* *** .:: ::       .:
         human ILGYTEDEVVSDDFNGSNHSSIFDAGAGIELNDT---FVKLVSWYDNEFGYSERVVDLMAHMASKE--   334
       Chicken ILGYTEDQVVSCDFNGDSHSSTFDAGAGIALNDH---FVKLVSWYDNEFGYSNRVVDLMVHMASKE--   332
          frog ILGYTQDQVVSTDFNGDTHSSIFDADAGIALNEN---FVKLVSWYDNECGYSNRVVDLVCHMASKE--   332
         trout YVGYTEYSVVSSDFIGDTHSSMFDAGAGISFNDN---FVKLISWYDNEFGYSHRVADLLLYMHFKE--   335
       lobster FLGYTEDDVVSSDFIGDNRSSIFDAKAGIQLSKT---FVKVVSWYDNEFGYSQRVIDLLKHMQKVDSA   333
           Fly ILGYTDEEVVSTDFLSDTHSSVFDAKAGISLNDK---FVKLISWYDNEFGYSNRVIDLIKYMQSKD--   332
          corn IMGYVEEDLVSTDFLGDSRSSIFDAKAGIALNDH---FVKLVSWYDNEWGYSNRVVDLIRHMFKTQ--   337
Dinoflagellate VLDWTDEEVVSTDFVSCKSSSIFDVGAGISLNDN---FVKLVSWYDNEWGYSNRLVDLAIYMAKKDG-   342
        E_coli VLGYTEDDVVSTDFNGEVCTSVFDAKAGIALNDN---FVKLVSWYDNETGYSNKVLDLIAHISK----   330
    Haloarcula VLGVTSDDVVSSDILGDPYSTQVDLQSTNVVSG----MTKILTWYDNEYGFSNRMLDVAEYITE----   334
   Helicobacter IIGIDSKQGVSSDFIGNPHSVVFAPDLSYIVNGN---MARVMAWCDNEWGYANRLLDMAQFISS----   338
       Aquifex ILQYCEDPIVSTDIVGNPHSAIFDAPLTQVIDN----LVHIAAWYDNEWGYSCRLRDLVIYLAERGL-   342
   Clostridium SFGYTEDEIVSADVVGISYGSLFDATLTKIVDVDGSQLVKTVSWYDNEMSYTSQLVRTLEYFAKIAK-   334
      Bacillus ILGYSEEPLVSGDYNGNKNSSTIDALSTMVMEG--S-MVKVISWYDNESGYSNRVVDLAAYIAKKGL-   334
         ruler .......310.......320.......330.......340.......350.......360........
```

Molecular Knowledge in G3PD

Referring to the alignment on the previous page, notice that some columns have an * placed above them. In these columns, all of the amino acids are identical. All of these columns contribute to knowledge. The other columns must be manually inspected. Columns that only contain amino acids from the same group (see chapter 4) also contain knowledge. The 64 columns that meet these criteria show up as rows in table 14.1.

To calculate the molecular knowledge using the genetic code, the amino acid group in each column is identified, and assigned a representative number of codons. For example, group 1 is composed of several amino acids that do not like water. This group includes alanine, methionine, leucine, isoleucine, and valine. Referring to table 4.1, alanine and valine are each specified by 4 codons, methionine is specified by 1, leucine is specified by 6, and isoleucine is specified by 3. Thus, the total number of codons that can code for a group 1 amino acid is 18. Thus, a group 1 column in the alignment (or the corresponding row in table 14.1) must contribute 3.32 x log (64/18) bits or 1.8 bits. The knowledge for the other groups is calculated in the same manner. For group definitions, see chapter 4, page 74.

Group Knowledge with Genetic Code
group 1 (leu, ile, val, ala, met) = 1.8 bits
group 2 (try, phe, trp)= 3.67 bits
group 3 (asp, glu)= 4 bits
group 4 (his, arg, lys)= 2.67 bits
group 5 (asn, gln)= 4 bits
group 6 (ser, thr)= 2.67 bits
group 7 (gly)= 4 bits
group 8 (pro)= 4 bits
group 9 (cys)= 5 bits

The procedure outlined above is then repeated using the primordial soup not the genetic code to determine knowledge. That is table 5.1 specifies the number of blocks for each group as explained in chapter 5. So from table 5.1, group 5 (asn and gln) contributes 400 + 400 = 800 blocks. Because there are 3,520,880 blocks in the truck (figure 5.1), the information is 3.32 x log (3,520,880/800) = 12.1 bits.

Group Knowledge with the Primordial Soup
group 1 (leu, ile, val, ala, met) = 3.1 bits
group 2 (try, phe, trp) = 11.5 bits
group 3 (asp, glu)= 7.4 bits
group 4 (his, arg, lys) = 14.8 bits
group 5 (asn, gln)= 12.1 bits
group 6 (ser, thr)= 10.2 bits
group 7 (gly)= 3.0 bits
group 8 (pro)= 12.2 bits
group 9 (cys)= 13.1 bits

Given these results, molecular knowledge is easy to calculate. Simply identify each column in the alignment that contains only amino acids from one of the predefined groups. If the gene or protein is theorized to have evolved after life exists, assign bits to each column using the knowledge with the genetic code. If the gene or protein is theorized to have evolved before the genetic code, assign bits based on knowledge with the primordial soup.

So referring to the alignment on page 225, column 11 is all G (which stands for glycine). So this column falls into group 7. It is assigned 3 bits with the primordial soup and 4 bits with the genetic code. Table 14.1 repeats this calculation for all positions that contain knowledge.

Table 14.1 GP3D Molecular Knowledge

Position	Molecular Knowledge Primordial Soup	Molecular Knowledge Genetic Code
11	3.0	4.00
12	11.5	3.68
13	3.0	4.00
15	3.1	1.83
16	3.0	4.00
17	14.8	2.68
21	14.8	2.68
38	12.1	4.00
39	7.40	4.00
54	7.40	4.00
94	12.2	4.00
99	11.5	3.68
109	7.40	4.00
111	10.2	2.68
112	3.0	4.00
114	11.5	3.68
120	3.1	1.83
123	14.8	2.68
128	3.1	1.83
131	3.1	1.83
134	10.2	2.68
147	3.1	1.83
150	3.1	1.83
151	12.1	4.00
164	10.2	2.68
166	3.1	1.83
167	10.2	2.68
168	13.1	5.00
169	10.2	2.68
171	12.1	4.00
173	3.1	1.83
175	12.2	4.00
178	14.8	2.68
179	3.1	1.83
184	11.5	3.68
186	3.1	1.83
195	14.8	2.68
198	10.2	2.68
201	12.1	4.00
205	7.40	4.00

214	14.8	2.68
219	3.1	1.83
226	3.1	1.83
227	3.1	1.83
228	12.2	4.00
231	10.2	2.68
232	3.0	4.00
233	3.1	1.83
240	3.1	1.83
242	12.2	4.00
244	3.1	1.83
246	3.0	4.00
247	14.8	2.68
248	3.1	1.83
250	3.0	4.00
254	14.8	2.68
255	3.1	1.83
256	12.2	4.00
286	3.1	1.83
310	3.1	1.83
311	10.2	2.68
313	7.40	4.00
340	14.8	2.68
351	11.5	3.68
total	**515**	**190**

Only the last row of table 14.1 is important. It shows that GP3D contains 515 bits of knowledge with the primordial soup and 190 bits with the genetic code. The odds of this protein arising by chance are 1 in 1.4 x 10^{155} with the soup and 1.1 in 10^{57} with the genetic code. Also from the table, the average molecular knowledge per amino acid may be deduced. Since G3PD contains 330 amino acids, the average molecular knowledge per amino acid is 1.56 bits with the soup and 0.57 with the genetic code (515/330 and 190/330 respectively).

Correction for Primordial Information

When using the genetic code, the machinery inside the cell ensures that an amino acid used by life is always placed in the protein. Even if the DNA changes randomly with no restraints, this is true. So amino acids with the genetic code are effectively free. The knowledge calculated in the previous section using the genetic code is correct.

The same cannot be said for the primordial soup. Even positions that allow all 20 amino acids used by life still contribute 2 bits (see chapter 5). This primordial information must be added to the knowledge calculated in the previous analysis. On average, G3PD contains 330 amino acids. Only 64 of these contributed to molecular knowledge. Thus, 330 - 64 = 266 amino acids that contribute 2 bits each need to taken into consideration.

Total Knowledge = molecular knowledge + primordial information = 266 amino acids x 2 bits + 515 bits = 1,047 bits.

Corrected average knowledge = 1,047/330 = 3.16 bits per amino acid.

The Average Molecular Knowledge Per Bit

The results arrived at here are very important. Because G3PD is a medium sized ancient enzyme, there is good reason to believe that its average knowledge per amino acid can be applied to other enzymes.

That is if one wants to calculate the molecular knowledge of a particular enzyme evolving in the primordial soup, the calculation is now trivial. If the enzyme in question contains 1,000 amino acids, simply multiple 1,000 x 3.16 bits per amino acid = 3,160 bits. The odds of the protein evolving in the soup are 1 in 2^{3160} or 1 in 1.8×10^{951}. Likewise, with the genetic code, the knowledge in a 1,000 amino acid enzyme is predicted to be 1,000 amino acids x 0.57 bits per amino acid = 570 bits.

The accuracy of the number so calculated will be very good for enzymes that show about the same level of conservation as G3PD.

To test this method, the molecular knowledge of the first enzyme in adenine synthesis should contain 290 bits of knowledge with the genetic code (510 amino acids x 0.57 bits per amino acid = 290). A direct calculation of knowledge using the same techniques that were applied to G3PD results in 303 bits. This is only a 4% error. So at least in this case, the procedure does seem to work quite well.

Perhaps an even better approach is to scale the results of the above calculation by the number of columns that contain knowledge. This technique may prove useful when more resolution is required. For the purpose of this book, the number of amino acids multiplied by either 0.57 or 3.16 is adequate.

Odds of ATP and Adenine Synthesis Evolving Concurrently

In chapter 13, it was shown that life requires 11 enzymes with 4,527 amino acids to synthesize adenine. But without ATP, these enzymes cannot function and offer no selective advantage, so the true number is 4527 + 335 = 4862.

To calculate the total molecular knowledge, multiply the number of amino acids by the 3.16 bits per amino acid calculated in this chapter.

Total Molecular Knowledge = 4862 X 3.16 = 15, 364 bits.

Or alternatively, if one assumes the genetic code is already in place:

Total Molecular Knowledge = 4862 X 0.57 = 2771 bits.

These results should be compared to those on page 213. The results on page 213 calculate knowledge using a completely different method and yet arrive at approximately the same numbers.

G3PD is shown below. The black and gray amino acids are highly conserved. Notice that most conserved amino acids are clustered on one side of the molecule. The cartoon version of G3PD is shown on the back cover (bottom right).

Figure 14.8: G3PD - Front and Back Views

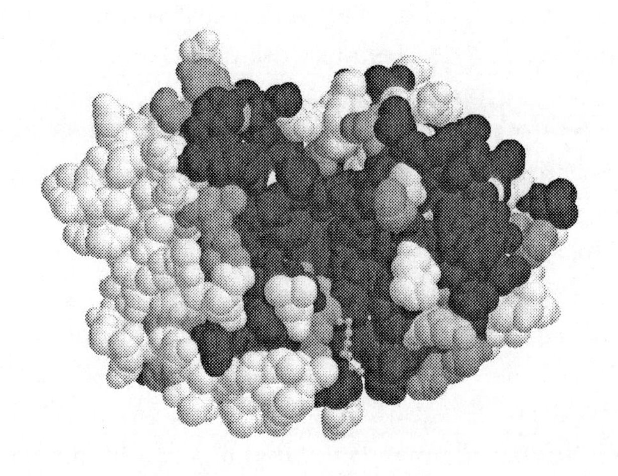

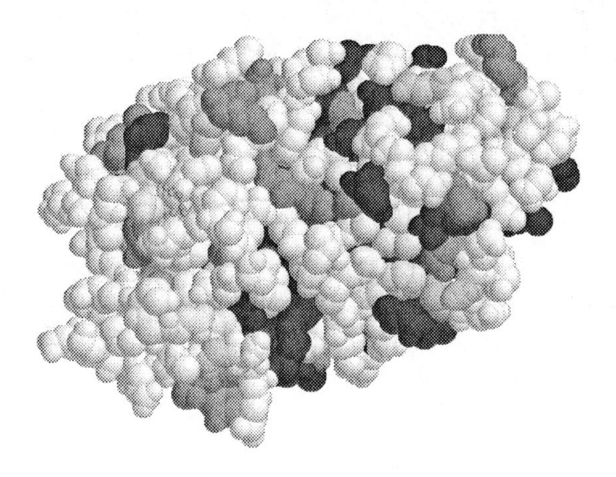

Does This Approach Generate False Knowledge?

As a control, 14 random sequences were generated using a tool designed to generate random sequences:

http://ca.expasy.org/cgi-bin/get-sprot-entry?RND21334

The length of each sequence was 500 amino acids. The sequences were then aligned using the same parameters that were used for G3PD. Of the 500 positions only one met the required criteria to be included in molecular knowledge calculations. The procedure was repeated several times, and the results were consistent. The technique described here usually filters 499 of the 500 columns in the alignment. Usually, only one column scores well. The only position that scored well in one example was all glycine. So this control experiment only creates 4 bits of information (with the genetic code).

Furthermore, when using the genetic code to calculate knowledge and information, every single position contributes some molecular knowledge - the knowledge to not terminate the growing peptide chain. Of the 64 DNA codons, 3 terminate the chain. Thus, at each position only 61 codons can be observed and 64 are possible. This is a form of both knowledge and information. This effect more than offsets the 4 bits artificially created above.

Perhaps more importantly, the technique described in this chapter filters out many positions that contain knowledge, and these positions are assigned 0 bits even though they contain quite a bit of knowledge (266 of the 330 sites in G3PD are filtered). So this procedures always underestimates the true knowledge.

ATP Synthase

ATP synthase is an incredible enzyme. It is the smallest rotary motor in the world. The protons moved across the cell membrane by the electron donor/acceptor/dehydrogenase complex (figure 14.6) serve as the energy source for ATP synthase.

ATP synthase lets these proteins flow back to the other side of the cell membrane, and this powers a small rotary motor imbedded inside the membrane and causes it to spin. The spinning portion called the rotor has a stalk attached to it. The stalk is not straight but rather curved. Because other peptide chains surround the stalk, as the stalk spins, it forces these surrounding peptides to move. This allows these surrounding peptides to create ATP from ADP.

ATP synthase was one of the first enzymes because it is absolutely necessary for many of the organisms that are thought to have existed on the primitive earth. All of the bacteria that oxidize non-organic chemicals to obtain energy use ATP synthase to make ATP.

The enzyme is composed of 8 distinct peptide chains. If any one of the chains is missing, the enzyme does not function. So ATP synthase is an irreducibly complex system. The subunits, their amino acid number, information and knowledge are shown in table 14.2. The operation of ATP synthase is illustrated in figure 14.9. The table and the figure together explain why this was not the first protein to synthesize ATP. If a system involving ATP synthase is required for the origin of life, then it will never get off the ground. The protein is too complex and contains too much knowledge.

Table 14.2: Information and Knowledge in ATP Synthase

Subunit (see figure 14.9)	amino acids in subunit	Knowledge (Soup)	Knowledge (Code)
A	240	758	137
B	160	506	91
C	71	224	41
Alpha	510	1611	291
Beta	480	1516	274
Delta	180	568	103
epsilon	133	420	76
gamma	288	910	164
total		6516 bits	1175 bits

This table assumes that the knowledge per amino acid is similar to that of G3PD. Thus, column 3 obtains molecular knowledge by multiplying the number of amino acids by 3.16 (primordial soup), and column 4 obtains molecular knowledge (genetic code) by multiplying the number of amino acids by 0.57. The last row is the sum of each column.

For this table to be correct all of the subunits must evolve independently, and this probably did not happen, because the sequence of the alpha chain is very similar to the beta chain. One of these two chains, probably evolved by gene duplication. Therefore information theory can still assign a number of bits, but these bits can no longer be related back to a probability because the knowledge is redundant.

To solve this issue, assume that the alpha chain is redundant knowledge and that it should not be included in the molecular knowledge calculation. This results in a molecular knowledge of 4905 bits (soup) and 884 bits (genetic code).

The chance of this protein evolving does not depend on whether or not the genetic code is in place. It is just too complex. The odds with the genetic code are given as follows: 1 in 2^{884} or 1 time in 1.28×10^{266} tries. Even without the math a closer inspection of figure 14.9, explains why the odds are so poor. ATP synthase is a single protein, and its very existence implies design.

Figure 14.9: ATP Synthase

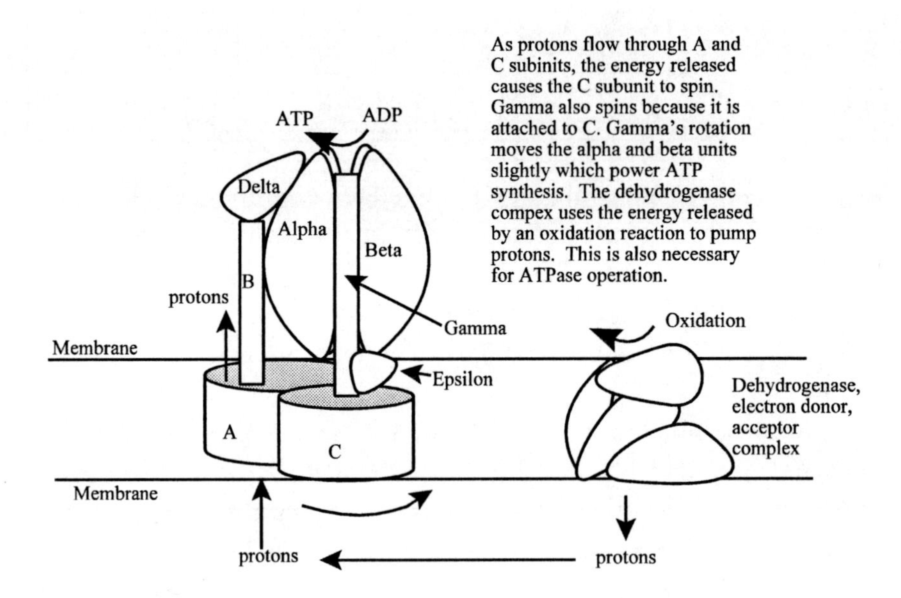

Conclusion:

The examples offered in this book are representative of the challenges that life faced at or immediately after its origin. Other systems of enzymes found in life are shared by bacteria and higher animals. For example, photosynthesis is carried out in both bacteria and in plants. The conclusion is that many if not most of these genes had to emerge right at the base of the tree of life. For purely logistic reasons, many of these genes were required by the first living organisms.

Any system of chemicals that does not know how to live is not a living organism, and this means that self replicating molecules only exist in text books. They will never be found in the lab, because the second law forbids their existence. In the end, there is no escape from the inevitable conclusion. The first living cell needed to emerge all at once, and chemical evolution cannot explain such a miracle.

On final idea needs to be considered. Many astronomers have suggested that the vast number of stars coupled with the extreme age of the universe can explain the origin of life, but these scientists never support their claims with statistics. They do not even bother with a single mathematical calculation. They state as a fact that life must exist on other planets because of the size and age of the universe. The next chapter will explore this claim.

References:

1) Watson et al. Molecular Biology of the Gene, 5th edition, 2004.
2) Stryer, Biochemistry, 1988.
3) Madigan et al., Brock Biology of Microorganisms, ninth edition, 2000.
4) Hutcheon et al., "Energy-driven subunit rotation at the interface between subunit a and the c oligomer in the Fo sector of Escherichia coli ATP synthase," PNAS, vol 99.
5) Zhou et al.,"Subunit rotation in Escherichia coli FoF1-ATP synthase during oxidative phosphorylation," PNAS, vol 94, 1997.

Part 4: Time, Natural selection, and Gene Duplication

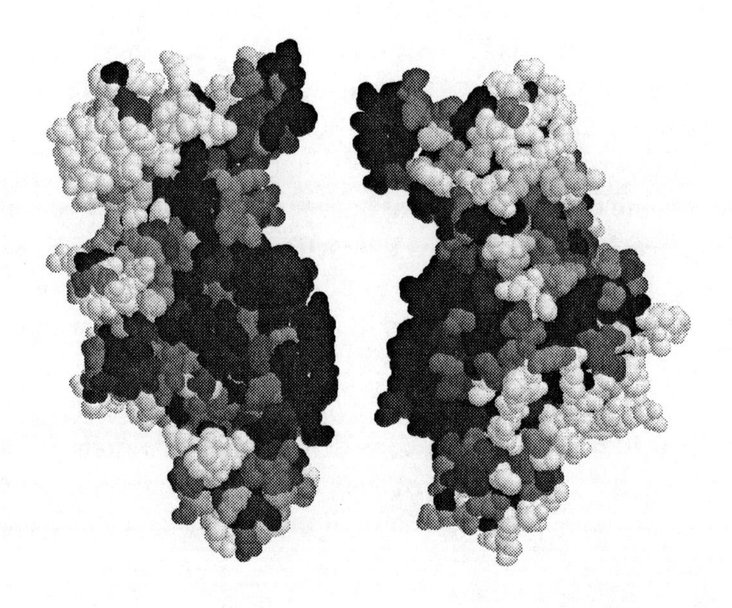

ATP Synthase alpha and beta subunits: Black and gray represent regions of high knowledge. White and light gray are regions of low knowledge

Chapter 15: The Effect of Time on Evolution

After life exists, knowledge can evolve, and this evolution is facilitated by both large populations and time. This chapter will show that when large populations are given several billion years to evolve, the knowledge that they create is very limited and often cannot even explain the origin of simple proteins.

How Does Time Factor Into the Equation?

If the odds of an event happening are 1 in 100, then with 10 tries the odds are approximately 1 in 10. Since every try takes time, time helps.

Consider a trapped scientist with a 12 word combination. If the scientist enters 12 words every 30 seconds, then he will enter approximately 1 million combinations over the course of a year. If the scientist lives for 5 billion years, and his basket contains 20 blocks labeled with words, then the odds that the scientist will open the door are 6 times in 10 tries. So in this case, time will most likely solve the problem by allowing chance to overcome a step in knowledge. Each word contains 4.32 bits of information (see chapter 1). So the 12 word combination contains 52 bits of knowledge, 12 x 4.32 =52. Thus, given 5 billion years and 1 million tries a year chance can probably overcome a 52 bit step in knowledge. The door in figure 15.1 is shown open because the scientist eventually enters the correct combination.

15.1: A Barrier That Chance Can Overcome

Now suppose that the door's combination is 21 words. This combination contains 21 x 4.32 = 91 bits of knowledge. The odds of opening the door are now 1 in 2.5 x 10^{27}. After 10 billion years with 1 million tries a year, the odds that the scientist will open the door improve to 1 time in 250 billion tries. The door stays shut.

Figure 15.2: A Barrier That Chance Cannot Overcome

Time can be represented as a growing tree (figure 15.3). The tree grows very slowly. After 5 billion years, the tree is almost 52 bits high. This allows the scientist to climb the tree and jump onto the ledge.

Figure 15.3: Time Represented by a Tree

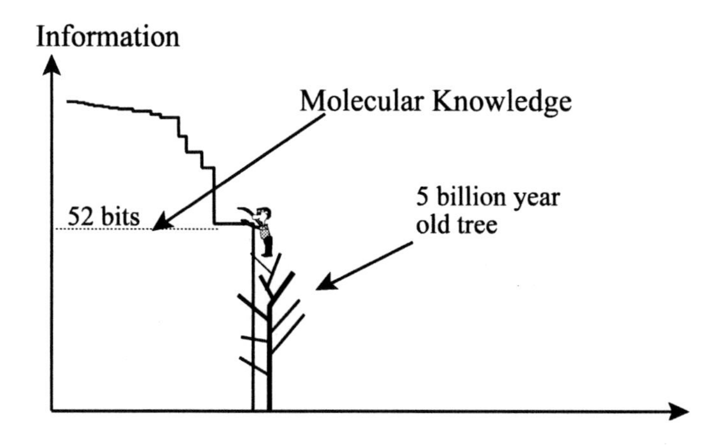

How Fast Does the Tree Grow?

Suppose that a scientist is given three dice and told to roll them until he throws triple fives. The odds that he will throw triple fives on the first roll are 1 in 216 (the dice have 6 x 6 x 6 = 216 possible outcomes, and only one is triple fives). What are the odds when the scientist throws the three dice twice? Many readers may think that the odds double. But this is only an approximation, and the approximation is only accurate if the odds are poor. The equation required to calculate the odds is as follows: odds of triple fives = $1 - (215/216)^{\text{number of rolls}}$. So with one roll the odds are $1 - (215/216)^1 = 1/216$ or 1 in 216. The odds with 2 rolls are $1-(215/216)^2 = 1/108.25$ or 1 time in 108.25 tries. Notice that the odds did not quite double.

Rolls	Probability	Odds
1	0.46%	1 in 216
2	0.92%	1 in 108.25
4	1.84%	1 in 54.4
8	3.65%	1 in 27.4
16	7.2%	1 in 14
32	13.8%	1 in 7.2
64	25.7%	1 in 3.9
128	44.8%	1 in 2.2
256	69.5%	1 in 1.4
512	90.7%	1 in 1.1
1024	99.1%	1 in 1.01

Figure 15.4 uses a bar to represent the probability of rolling triple fives. The numbers along the bottom represent the number of tries. When the bars are short, each successive bar is almost twice as high as its predecessor. Once the probability is greater than ten percent, doubling the tries no longer doubles the probability. The probability for success will never be equal to 100%, but after 1024 tries, it is very close.

Figure 15.4: Probability of Rolling Triple Fives

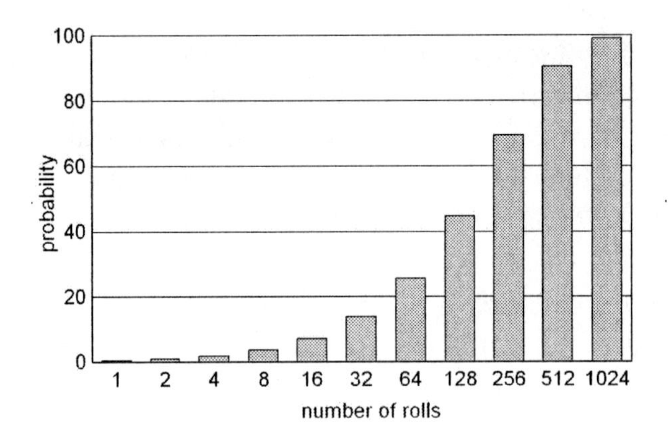

With quite a bit of mathematical manipulation, figure 15.4 can be converted into figure 15.5.

Figure 15.5: A Growing Tree Helps the Scientist Climb the Wall

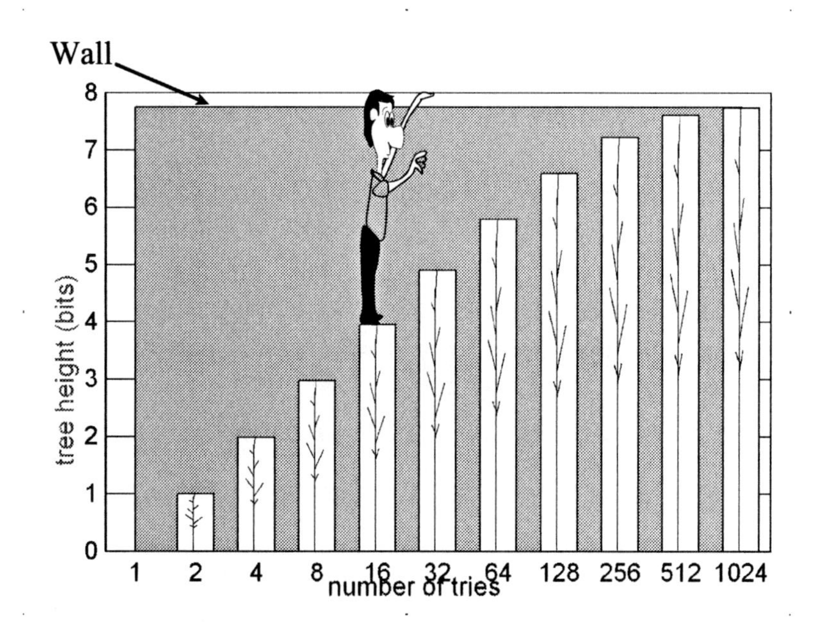

In figure 15.5, each bit represents one foot. Because rolling triple fives corresponds to a 1 in 216 chance, the initial height of the wall in figure 15.5 is 7.555 bits (information = 3.32xlog(216/1) = 7.75 bits) or 7.75 feet.

After two rolls, the height of the wall is given as follows: information = 3.32 x log(108.25/1) = 6.75 bits. Rather than shrink the wall, which is hard to draw, figure 15.5 shows the scientists standing on a tree. The height of the tree is the initial height of the wall minus the new height of the wall. Thus, after two rolls the tree is 7.75- 6.75 = 1 foot high.

After 16 rolls, the odds improve to 1 in 14. Thus, the new height of the wall is equal to 3.32 x log (14/1) = 3.8 bits or 3.8 feet. To compensate the tree must be 4 feet tall (7.75 - 3.8 = 3.95).

The scientist is standing on the tree that corresponds to 16 rolls. He only has a 1 in 14 chance of climbing the wall. After 1024 rolls, there is almost no chance that he will not be able to climb over the wall.

To relate this example to evolution, each roll of the dice corresponds to a try, and each try corresponds to a reproductive event. So how fast the tree grows depends on reproductive rates. Animals that have large populations accumulate many more tries than those with small populations. Animals that reproduce slowly like elephants will accumulate fewer tries than animals that reproduce quickly like rabbits. Since it takes time to accumulate tries, the number of tries can easily be converted into years. If the scientist rolls the dice once a year, then the x-axis in figure 15.5 can be written in years, and it will take the tree that the scientist is standing on 16 years to grow.

Upper Limit on Number of Tries

The only opportunity for chance to accumulate tries is during reproduction. Thus, the number of tries that any animal or plant accumulates each year is proportional to how many offspring it produces.

The most abundant, fastest reproducing organisms accumulate the most tries. The unquestionable leaders are bacteria, and the numbers are staggering. For every insect on the planet there are 500 billion bacteria. For every star in the universe, there are 10 million bacteria. Furthermore, when conditions are optimal one bacterium can split into two bacteria in a matter of minutes. One study estimates that 1.7×10^{30} bacteria are born each year.[1]

The next chapter will show that existing genes are not free to evolve into new genes. Thus, the knowledge in genes must be duplicated before its evolution can yield a new gene.

Assume that at any given time 20% of all the DNA in bacteria is duplicate information. On average, bacteria contain 2,000 genes. This means that 400 genes in each bacterium are free to evolve outside of the constraints imposed by natural selection (see chapter 16).

Assume that one mutation is created for every 10,000 genes replicated (this mutation rate is much higher than it is today). With this assumption, every 25 replication events result in a duplicate gene mutation, and bacteria accumulate 6.8×10^{28} tries a year ($1.7 \times 10^{30}/25$). Over 1 billion years, they accumulate 6.8×10^{37} tries. This is very significant because it gives chance an opportunity to create a substantial amount of knowledge. With this many tries, it is quite likely that bacteria can open a door with a 29-word combination. In other words, given a billion years, and a massive army of bacteria, chance very well may create 126 bits of knowledge. Since some proteins are functional with less than 126 bits of knowledge, it is quite probable that these proteins evolved by chance.

More importantly, bacteria can open the equivalent of a door with a 22-word combination in a single year. So for bacteria, the amount of knowledge created in a single year is almost as great as that created in a billion years. This happens because bacteria accumulate tries through large populations and fast reproductive rates, and these two numbers are very large compared to 1 billion years.

This also means that the idea that science cannot observe large changes because of time constraints on scientific experiments is nothing more than a popular myth. Why this myth may be true for evolution in mammals, it is certainly not true for bacteria.

Now consider an animal with a smaller population. Assume that people will populate the earth for the next 50 million years, and that over this time on average, 300 million people will be born each year. Over the next 50 million years, 15,000 trillion people will be born. Because people have quite a bit more DNA than bacteria, assume that everyone has 5,000 duplicate genes, but the accuracy of DNA replication is now better. Suppose that a mistake is made on average every time 100,000 genes are copied. Thus, evolution accumulates 1 try with every 20 births (100,000/5,000 =20). The total number of tries is thus 75 trillion (15,000 trillion/20). With this many tries, it is quite probable that chance will be able to find the combination for a 10 word door, and this corresponds to 43 bits of knowledge. So evolution in people is not very productive.

The population of most mammals is so small that they simply cannot accumulate enough tries to create a new protein.

Figure 15.6 shows how the trees for different species may help them climb over a wall of knowledge. The height of each tree depends on two factors 1) its age and 2) the number of tries accumulated each year. In general, the number of offspring produced by a species is inversely proportional to its physical size. Mice have more offspring than elephants. So each species requires its own tree.

Each tree in figure 15.6 starts growing today and is allowed to grow for 50 million years. Thus, each tree is 50 million years old. The tree helping bacteria is very tall because bacteria accumulate the most tries. The tree helping elephants is short because elephants do not produce many offspring.

Notice that in figure 15.6, there is a scientist who lives before life exists and that he does not have a tree. Synthesizing complex chemicals before life is so difficult that given 5 billion years, the number of tries will probably still equal zero. But even if the number of tries equals 100,000, then this would correspond to the prebiotic scientist standing on a rather tall blade of grass. Time only becomes important after life exists.

Figure 15.6: A Tree for Every Species

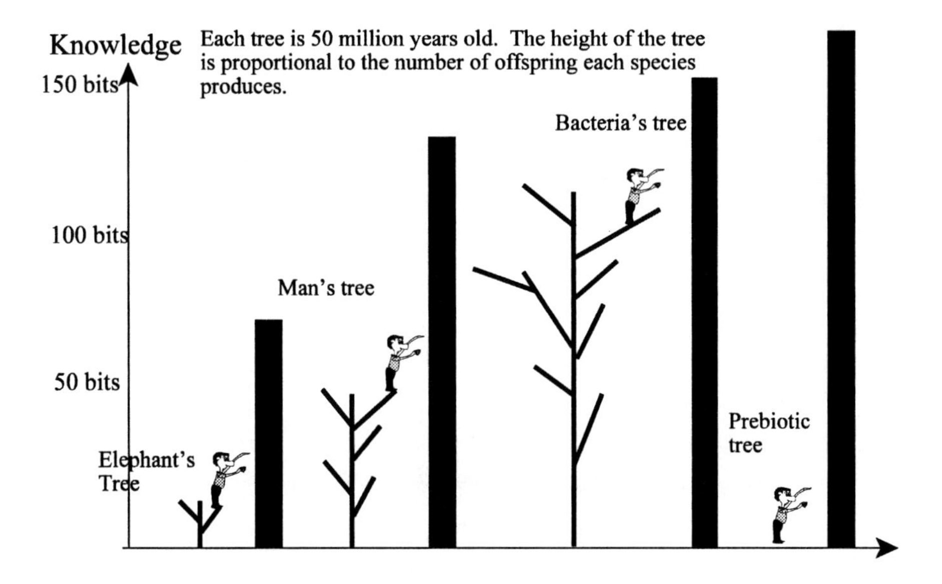

The Trees Help, but . . .

With the trees, the origin of a few proteins can be attributed to chance. Nevertheless, the vast majority of proteins have too much knowledge (chapters 13 and 14). The protein ATP synthase is an example of a single protein from which design may be inferred. Not only was it one of the first proteins, but it contains quite a bit of knowledge, 884 bits. As far as the trees are concerned, 884 bits is an unreachable target. Furthermore, the design inference never has to rest on a single protein. A much stronger case for design can be made by selecting a set of several proteins that perform a specific function. If the chosen system is irreducibly complex, then all of the protein must evolve before a selective advantage is realized. The molecular knowledge of these systems is so great that relying on chance to explain their origin is irrational. The best examples are the enzymes responsible for adenine synthesis and ATP synthesis, 2771 bits. The trees cannot grow this tall. The scientist can never climb a wall this tall.

The Origin of Life

All origin of life theories inevitably run into the same problem - zero tries.

Before bacteria and insects, chance just does not get many tries (if any). Prebiotic chemistry cannot rely on $5x10^{30}$ bacteria accumulating $6.8x10^{28}$ tries every year. Before life exists, there are no growing trees to help the scientist climb over the walls of knowledge.

Biological evolution accumulates an incomprehensible number of tries. The same cannot be said for chemical evolution. The experimental evidence gathered over the past 50 years suggests that very few if any biological precursors would have existed on the primitive earth. The trapped scientist cannot open the door unless he types in words. His chance of finding the combination improves with the number of tries, but his chance of opening any door with zero tries is zero, so even if the combination is a single word, he will not find it. Zero tries are so problematic, because the tree in figure 15.6 is not just small, it no longer exists. Figure 15.7 shows the effect of zero tries. The scientist can never open the door.

Figure 15.7: The Origin of Life

Reference:

1) Whitman, Coleman, Wiebe, "Prokaryotes: The Unseen Majority," PNAS, 95:6578-65-83, 1998.

Chapter 16: Natural Selection Preserves Existing Genes

This chapter will use the trapped scientist to show why natural selection preserves existing genes.

Suppose the trapped scientist is now in a two-story building. The computer starts with a message already in it, and this message contains the knowledge to open all of the doors on the first story. The combination is dog-computer-cat-cat-bike-book-book-run-man-sun-dog-dog. The scientist is given two baskets. One contains 20 blocks labeled with words, and the other contains 12 blocks labeled with the numbers 1 through 12 (figure 16.1).

Figure 16.1: Natural Selection Preserves Existing Genes

The scientist is instructed to draw one block from each basket. He is to use the number that he draws to locate a position in the door's combination, and he is to change the existing word at that position to the new word which he draws. For example, on the first try, the scientist draws the number *12* and the word *cat*. The original combination has the word *dog* at position 12. So the scientist replaces this word with the word *cat*. When he makes this change, the last door on the first floor slams shut because its combination is no longer correct. The scientist climbs down the ladder and realizes that he is trapped. He becomes very agitated. He changes the word *ca*t back to *dog*, and the door opens.

He leaves and refuses to participate in any further experiments. He has no desire to be trapped in the room, and his refusal to participate preserves the combination that opens the first floor doors.

In this example, the combination that opens the doors on the first floor represents a gene, and the scientist represents natural selection. The scientist preserves this existing gene by refusing to participate in the experiment.

If protein A is represented by the bottom doors, then this protein is preserved by natural selection. If the upper doors represent protein B, then this protein will not evolve. The reason is simple. The preservation of the combination that opens the first story doors prevents the top floor combination from being found.

This Simple Example Shows that Evolution Does not Work Quite Like Darwin Imagined

When Darwin introduced the theory of evolution, he envisioned everything being guided by natural selection. The following quote conveys his thinking:

“If it could be demonstrated that any complex organ existed which could not possibly have been formed by numerous, successive, slight modifications, my theory would absolutely break down.” - Charles Darwin

Darwin should be applauded for this particular statement. This quote is right on target. Darwin wanted to explain evolution with small continuous steps. He took a very simplistic approach. The legs of reptiles can gradually over many generations evolve into wings. The fins of fish can gradually evolve into legs for reptiles. He thought all of these changes were guided by natural selection. This simplistic approach is still taught in high school and college biology. Whether true or not, it is a great way to teach evolution because it minimizes the role of chance and makes the theory seem more reasonable.

Some biologists do not realize that useful information (any door’s combination) is locked by natural selection. The fins of a fish are not free to evolve into the legs of a reptile. The legs of a reptile are not free to evolve into the wings of a bird, and it is natural selection that makes this a certainty.

Natural Selection Reduces the Number of Tries

Earlier a tree was used to represent the passage of time. As the tree grows, the scientist is able to climb it, and this allows him to climb steps in knowledge. The effect of natural selection on the tree's growth is shown below. The tree is now much shorter; as a result, the steps that can be overcome by chance given several billion years are much smaller.

Figure 16.2: Natural Selection Reduces the Number of Tries

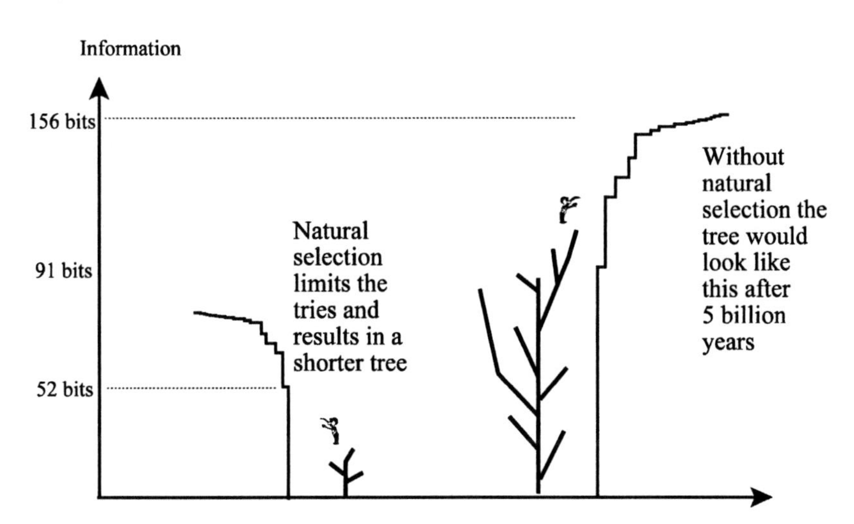

As radical as this concept may seem to some readers, it is not really in dispute. Very few molecular biologists would disagree with this particular point. Many have already stated it very clearly.

"We have no trouble understanding how natural selection can maintain a functional single-copy gene like globin or insulin. If the gene product is defective in any serious way, the organism producing it will be immediately subjected to a selective disadvantage; it will either die prematurely or produce fewer progeny that its unmutated siblings" - Molecular Biology of the Gene, Watson et al, 1987.

"As long as a particular function of an organism is under the control of a single gene locus, natural selection does not permit perpetuation of mutations which result in affecting the functionally critical site of a peptide chain specified by that locus. Hence, allelic mutations are incapable of changing the assigned function of genes." - Evolution by Gene Duplication, Ohno.

"Gene duplication must always precede the emergence of a new gene having a new function." - The Neutral Theory of Molecular Evolution, Kimura."

Small changes are cumulative only when they optimize an existing protein. Natural selection guides this optimization. Once optimized, the changes don't stop, but they are no longer cumulative because natural selection's role switches from one of optimization to one of preservation. Thus, large changes are not expected even if evolution is given millions of years to operate. One of the best examples of the preserving power of natural selection is insulin. Insulin in fish is almost identical to insulin in humans.

To summarize, natural selection prevents evolution from happening like Darwin envisioned. Darwin's small changes only optimize information. That is such changes can make a reptile's leg a better leg. These changes can improve the eyesight of a cat. In this chapter, such changes are represented by a series of doors on the same floor opening. Let the series of doors on the top story represent a bird's wing and the series of doors on the bottom represent a reptile's leg. Once the solution for the reptiles leg is found, the bird's wing will not evolve. Large changes need to be explained with another mechanism because the slow and gradual explanation proposed by Darwin does not work. Large changes must rely on information being duplicated. Duplication is very important because it frees information from the constraints imposed by natural selection.

Hopeful Monsters

Punctuated equilibrium is the theory that new animals and plants arise almost instantaneously. The theory was introduced to help explain the lack of transitional forms in the fossil record. Most fossils that are found are already fully evolved. For example, when a fossil of a dinosaur like T. Rex is found, it is very easy for paleontologists to identify it as T. Rex. If Darwin was right and changes are gradual, this is not what one should expect. A continuous line of fossils documenting the evolution of T. Rex is expected. The fossil record does not support slow gradual change.

Punctuated equilibrium is a theory of hopeful monsters. That is massive change due to numerous mutations in a few offspring give rise to new genes and proteins. The change is very quick (one generation), and it has nothing to do with intelligent design (naturalistic laws are assumed to be in control).

For example, imagine a population of fish whose fins are optimized for swimming. Because of natural selection, these fins must always stay optimized for swimming. But the lake that these fish are in is about to go dry. And fortuitously, a few of the offspring from one of the fish are born with very strange fins. These fins are no longer optimized for swimming, but they do allow the fish to walk better. Normally such fins would simply be eliminated from the population by natural selection. Nevertheless, when the lake dries up, these fish may provide an advantage if the fish can make it to another lake. So the change from fins optimized for swimming to those that enable the fish to kind of swim and sort of walk happens instantly (in a single generation).

Because of the preserving power of natural selection, this is really the only option for evolution. Chance creates knowledge in a single generation and a door opens.

Can Intelligent Design be Applied to the Evolution of Mammals?

After reading this chapter, some readers are probably thinking that the goal was to extend intelligent design to the evolution of higher animals and plants. While this may or may not be a valid argument, it was certainly not the goal of this chapter. The goal of this chapter was simply to show that evolution does not work like most people think it does.

Intelligent design is very difficult to apply to the evolution of mammals for several reasons. There is not a single gene in man that is not found in an ape. The techniques outlined in this book just do not work if one's goal is to show that an existing gene in a chimpanzee cannot change into a slightly modified version of the same gene in man.

To further complicate matters, most evolution in mammals appears to be the result of gene duplication and differential gene expression. That is existing information is duplicated and used in different ways. For example, suppose the sentence *I see a cat* is copied and then modified to say *I see the cat*. The subtle differences in these two sentences are much greater than the differences observed between the DNA in a man and a mouse. If information theory is used to calculate the odds of the second sentence evolving, then the calculation will be wrong. When duplications are involved, the probability for success depends on the path of evolution, and this is something that information theory cannot deal with effectively.

Perhaps the greatest difficulty lies with differential expression of the same genes. Most mammals share the same genes. In the evolutionary path from mice to men, few if any new genes were invented (Watson: 2004). The genes found in man that do not have a corresponding gene in mice are the result of simple gene duplications.

Differential expression of the same genes is a very rich source of variation. All of the cells in a man have the same DNA. Yet a muscle cell is very different from a skin cell or a liver cell, and this difference is due to differential expression of the same genes. In other words, while a liver cell may have the same genes as a muscle cell, the set of genes expressed in a liver cell is different than those in a muscle cell. Some genes are repressed in the liver cell compared to the muscle cell, and others are much more actively expressed in the liver. Science has identified regulatory genes called pattern genes. Mutations in these pattern genes can control which genes are expressed in specific tissues, and simple mutations in these pattern genes can give rise to new distinct traits like fruit flies with eyes on their legs. How all these regulatory paths influence variation is still poorly understood.

The variation observed in mammals does not require the invention of new genes. So the mathematical models described in the preceding chapters cannot be used to model mammalian evolution. The next chapter will explain why the problem is so hard to model.

When dealing with the origin of life and the evolution of the first genes and proteins, the answer to the question, is evolution possible, is easy to answer, and the answer is no. This is not the case for vertebrate evolution. Today, science is not even prepared to ask this question because nobody understands exactly what is required to change an ape into a man.

References:

1) Ohno, Evolution by Gene Duplication, Springer-verlag, 1970.
2) Watson et al. , Molecular Biology of the Gene, Cold Spring Harbor, 1987.
3) Lewin, Genes III, Wiley & Sons, 1987.
4) Kimura, The Neutral Theory of Molecular Evolution, Cambridge, 1886.
5) Gould S.J., and N. Eldredge, “Punctuated equilibria: the tempo and mode of evolution reconsidered,” 1977, Paleobiology.
6) Watson et al, Molecular Biology of the Gene, Fifth Edition, Cold Spring Harbor, 2004.

Chapter 17: Evolution by Duplication

In order for new proteins to evolve, a way around the preserving power of natural selection must be found. If information is first duplicated, then natural selection can be removed from the picture. Consider the trapped scientist again. This time he has two computers, two screens, and four baskets (figure 17.1).

At the start of the experiment, both computers already have the combination for the bottom doors loaded. The scientist is given two baskets on the second story. He is told to pick a block from each. One contains 20 blocks labeled with words, and the other contains 12 blocks numbered *1* through *12*.

Figure 17.1:Duplicate Information

As before, the number determines the position in the door's combination that needs to be changed. For example, suppose the scientist draws the number *1* and the word *plant*. He changes the first word on the upstairs computer to *plant* and nothing happens. He then climbs down the ladder and draws two blocks from the first story baskets. He draws the number *2* and the word *bike*, so he changes the second word on the downstairs computer to *bike*, and the first door slams shut. He becomes agitated because he is now trapped. He restores the downstairs computer to its original state, and the door opens. He decides that he is not going to touch the downstairs computer again. Thus, his decision preserves its combination.

He climbs the ladder and continues to change the combination on the computer upstairs by following the procedure outlined above. He eventually opens all of the doors (figure 17.2)

Figure 17.2: Duplicate Information All Doors Open

Figures 7.1 and 17.2 model gene duplication. The combination of the doors on the first story represents an existing gene. This gene is duplicated by placing its knowledge on the upstairs computer. The scientist is able to change the upstairs computer because natural selection does not preserve duplicate genes.

Protein Families

The amino acid sequence is often very similar for related proteins. Gene duplication explains how these proteins evolve. Consider the trapped scientist with two computers again. Initially, both computers contain the combination necessary to open the bottom door. The combination of the door on the first floor represents an existing gene. The combination of the closed door on the second story represents a duplicate gene that is in the process of evolving but does not yet confer a selective advantage (figure 17.3).

There is only one difference in the two combinations. The upstairs door uses the word *drink* instead of *coconut*. If he is lucky enough to draw the number *1* (1 in 12 chance) and the word *drink* (1 in 20 chance), the scientist will open the second story door on his first try. The odds that he will open this door on his first try are 1 in 240. Now imagine the scientist in a 241-story building (to model 240 duplicate genes). All of the additional floors start off in the same initial condition as the second story in figure 17.3. That is each computer starts with the combination for the door on the first floor, and each computer only needs its first word changed to the word *drink* in order for the door to open. The scientist starts on the top floor, draws two blocks, and if the door opens, he leaves. If it does not, he moves down a floor and tries again. Thus, he accumulates 240 attempts to open a door with 1 attempt on each story. He will probably find the combination. Many proteins have evolved in this manner, and because the odds are so good, design need not be inferred.

Figure 17.3: Protein families

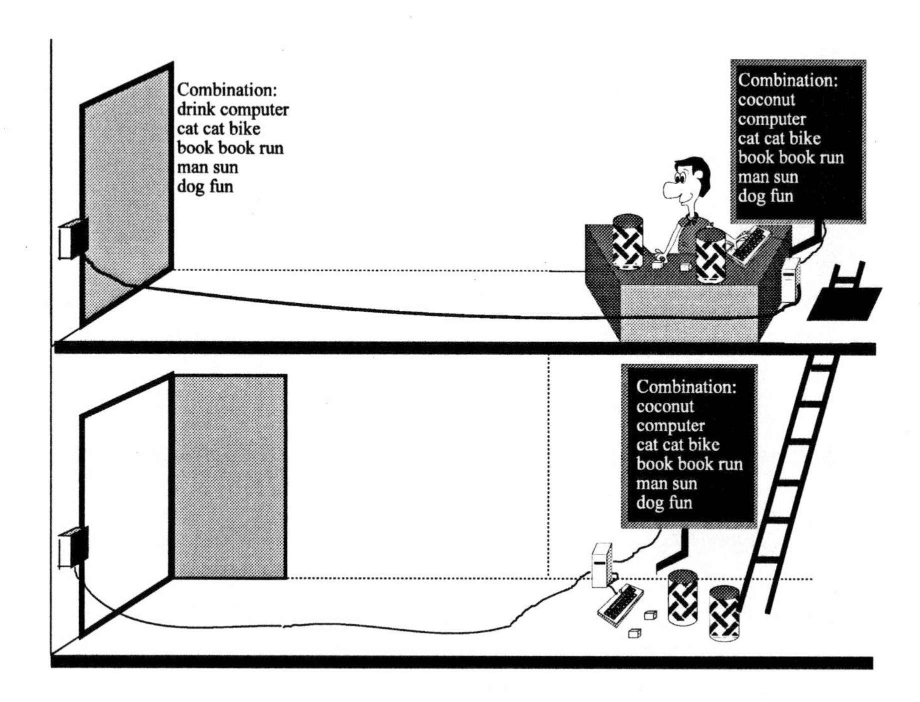

Protein Diversity

Duplication works well with similar proteins. It does not help when the proteins are very different. Consider the previous example again. Now the combination for the door downstairs has nothing in common with the combination upstairs. Starting with the combination for the downstairs door in this case is no better than starting with a random combination. Duplication will not help break the code for the top door (figure 17.4).

Figure 17.4: Duplicate Information Does Not Help In This Case

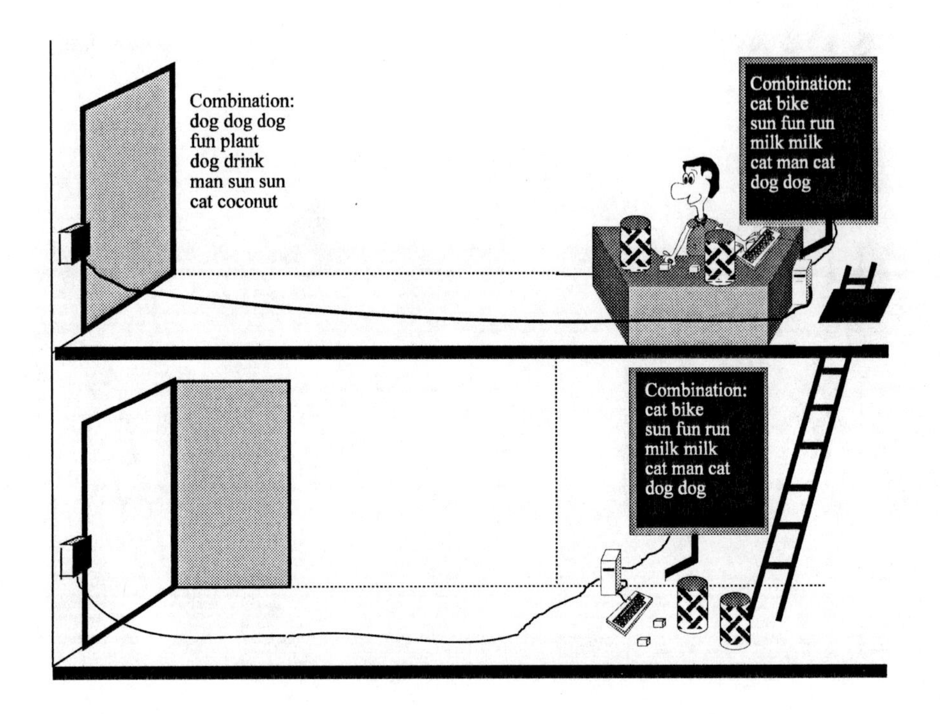

Figures 17.3 and 17.4 are the two extremes. One could envision a combination that only needs two words changed or perhaps three, but every time a change is necessary, the required mutation probably will not occur in the correct place, and as these undesirable mutations accumulate, they tend to destroy the protein. If a duplicate gene requires 5 or more specific mutations to evolve into a new protein with a new function, then its function will almost always be destroyed by undesirable mutations before the desired mutations take place. This limits the effectiveness of gene duplication. Figure 17.5 illustrates this concept.

Figure 17.5: Most Duplications Are Lost

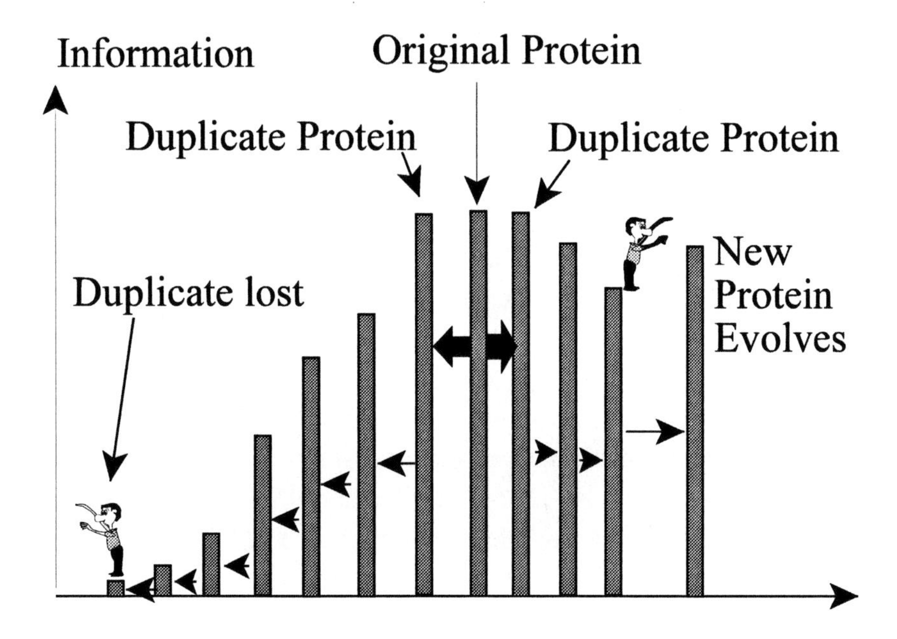

In figure 17.5, the original protein is duplicated twice (thick arrows). Mutations then change the amino acid sequences (thin arrows). Each mutation destroys information, so each wall becomes shorter as it mutates. If the new protein is similar to the old, enough knowledge will survive to allow the scientist to jump on the new wall (the path to the right). Otherwise all of the knowledge will be lost (path to the left). In mammals and other higher life forms, the lost duplicates give rise to what is called junk DNA. It is non-coding DNA that serves no purpose. The burden of carrying this extra DNA does not bother organisms with slow reproductive cycles. Lost duplicates also give rise to pseudogenes. These are genes that are very similar to their functional ancestors but do not encode a useful protein. Bacteria can reproduce in 10 minutes so any extra DNA is an extreme selective disadvantage. Bacteria only keep the genes that they need.

Duplicate Genes That Require Multiple Mutations

In a ground breaking paper, Behe and Snoke, published an article in Protein Science (April 2004) that at its very core asks the question that this book has been trying to answer from page one - namely is evolution possible?

In ths article, Behe and Snoke calculate the likelihood that multiple point mutations will create a new gene from a duplicate gene. The article is so suprising because the authors were able to freely ask if evolution is possible. While the article is very good, it is somewhat limited in scope in that it only considers point mutations.

The trapped scientist will now be used to show that the most likely mechanism for duplicate gene evolution involves information shuffling, and this is unfortunate because it greatly complicates the mathematical models. The probability of evolution becomes path dependent. So one must know the path by which a new gene evolved to calculate the probability. The goal here is not to introduce such a model, but rather explain some of the difficulties.

The trapped scientist is given a basket with 28 blocks labeled *A* through *Z*, one block is labeled with a space, and one with a period. As a starting point, the computer always begins with this message: The brown dog is a lab.

The scientist is given a second basket containing blocks labeled with the numbers *1* through *23*. The message and the corresponding numbers are shown below.

1	2	3	4	5	6	7	8	9	10	11	12	13	14	15	16	17	18	19	20	21	22	23
t	h	e		b	r	o	w	n		d	o	g		i	s		a		l	a	b	.

The scientist is to draw one letter and one number from each basket. He is to use the number to find the correct letter to replace. So if he draws a *10* and a *K*, he should replace the space between the word *brown* and *dog* with the letter *K*. If he spells the following sentence, the door will open, and he can leave: the black dog is a lab. A quick inspection reveals that the scientist might as well be starting with a random message. For him to spell this combination correctly, his first four draws must be confined to the numbers *6, 7, 8*,and *9*. Furthermore, when he draws a *6* he must also draw the letter *l*. When he draws the number *7*, he must draw the letter *a*. When he draws the number *8* he must also draw the letter *c*. And finally when he draws the number *9*, he must also draw the letter *k*. This is not going to happen. The odds are too poor.

The scientist begins to change the message, and after 20 changes, the computer automatically resets back to the original message. Each reset event corresponds to a new gene duplication. The scientist will likely be reset to the original 200 billion times before he actually manages to open the door. With enough time, this is a solvable problem, but it is extremely difficult.

So now consider another experiment in which instead of letters the scientist is given a basket with 30 words. All of the words that he needs to open the door are in the basket. The numbering system is now as follows:

1			2	3					4	5			6	7		8	9	10	11			12
t	h	e		b	r	o	w	n		d	o	g		i	s		a		l	a	b	.

On the first try, if the scientist pulls the word *black* (1 in 30 chance) with the number *3* (1 in 12 chance) the door opens. Before 400 resets, he will probably open the door. This is a much more plausible path for duplicate gene evolution when the gene requires more than one mutation to acquire a new function.

Protein Domains

The trapped scientist in the previous section will now be related to how duplicate genes may evolve. Changing letters corresponds to changing single amino acids in the duplicate protein (point mutations). This path is viable if and only if a few amino acid changes will create a new function.

Changing words corresponds to shuffling protein domains. A protein domain is a small section of a protein that serves some function. The function might be structural or it might be chemical. There are quite a few mechanisms in cells by which this shuffling can happen. For example, during meiosis (the process by which higher animals reproduce), the chromosomes align and exchange DNA. Sometimes this exchange does not happen as designed, and one chromosome may end up with extra DNA, and the other will lose DNA. This process can transfer a protein domain from one gene into the middle of another gene. It is just like changing the words in the trapped scientist example. There are many different mechanisms in cells that can transfer domains from one gene to another. Furthermore, the evidence that many proteins have evolved through this mechanism is compelling. Unrelated proteins often share the same domains. That is a domain with some function evolves once, and this domain appears in many distinct unrelated proteins. The trapped scientist example shows the power of this technique, and because the process is path dependent, information theory has a very difficult time modeling the probabilities associated with such evolution.

This is why this book confined its analysis to evolution that took place before and around the origin of life. Gene duplication and information shuffling must be ruled out to make the design inference compelling. If more scientists would attempt to ask if gene duplication and information shuffling can explain the emergence of new genes, then the answer would not prove so elusive. It is a very difficult problem, but it is solvable. Today, science assumes naturalistic laws are responsible. One day, science must take the next step and test this critical assumption.

Chapter 18: Alternatives to Intelligent Design

Science depends on two axioms, the naturalistic axiom and the observable axiom. Axioms are assumptions. They are supposed to be self evident, but in many cases they are not. Axioms cannot be proven. They are accepted on faith. Unfortunately, neither of the axioms on which science is based are self evident to everyone. Nevertheless, science needs these axioms to function properly.

The naturalistic axiom allows science to assume that everything can be explained with math, physics, and chemistry. The observable axiom states that man is capable of formulating laws and theories that describe nature from his observations and experiments.

Today, the problems associated with chemical evolution, the origin of life, and the evolution of the first genes and proteins have backed science into a corner, and science has no way to cope with these issues. The naturalistic axiom does not allow for the possibility of design, and the observable axiom suggests that science should be able to find solutions to these mysteries.

The previous chapters suggest that the naturalistic axiom is not valid. This chapter will consider the alternative. That is suppose that the observable axiom is not valid. The implications open the door to an endless number of possible solutions.

Once the observable axiom is dismissed, all scientific theories and laws are immediately called into question, and science cannot be sure of anything. Without the observable axiom, science becomes a useless academic exercise.

Science without the Observable Axiom

Many solutions to chemical evolution and the origin of life exist. The observable axiom does not allow science to consider these possibilities. But without the axiom, they must be considered.

- Man's observations may influence the results. That is chemicals may combine and form life in small puddles quite easily as long as nobody observes the process. If this is true, then the experiments conducted over the past 100 years are no longer relevant. That is spontaneous generation is quite common, but it only happens when scientists are not around.

- Perhaps, there are an infinite number of stars, and astronomers cannot see them. Maybe they are too far away. Maybe they exist in a parallel universe. Infinity has many nice properties that solve the problems associated with both chemical evolution and the origin of life. No matter how poor the odds, with an infinite number of tries, the solution will always be found.

- Perhaps, the world and the universe are artificially created programs running inside a powerful computer; as a result, every individual is just a computer program. The problems associated with evolution are merely computer program errors.

- Perhaps, matter and energy have some vital force that man cannot observe. This force causes matter to organize into life.

The list of possible solutions is only limited by the reader's imagination. While only four solutions are listed above, the number of solutions is endless. Not everyone will agree, but this author feels that the observable axiom is self evident, and as such, the possibilities listed above do not deserve serious consideration.

Consequences of the Observable Axiom

If the observable axiom is true, then the nature of how the designer created life is open to observation, and several key observations can easily be formulated from the scientific evidence.

- The designer was most active 3.5 billion years ago. The origin of life required quite a bit of help to get off the ground because many critical genes seem to be coincident with the origin of life, and the necessary biological molecules are not produced in abundance by nature.

- Once the first living cell was created, the designer seemed to allow bacteria to create variation and optimize new genes and proteins through the naturalistic process of evolution. This process continued for 3 billion years, and it relied heavily on the large populations and fast reproductive cycles of bacteria. Many new genes unrelated to previous genes may have been created by the designer during this time.

- Five hundred millions ago, it appears that the designer probably stepped in again and created most of the major biological phyla during the Cambrian explosion.

- It is very difficult to observe the influence of the designer in higher animals and plants because the processes are path dependent and not easily described by information theory. It is chemical evolution and the subsequent origin of life that make the design inference so compelling. Design arguments concerning higher life may be valid, but they are much harder to justify.

Appendix 1: Shannon Entropy and Information

In chapter 1, Shannon entropy is mentioned in passing. Shannon entropy is intimately tied to information theory. For readers who may read more about information theory elsewhere, this section will prove useful.

Information is transmitted in symbols. These symbols are most frequently letters. The English alphabet is a group of symbols. These symbols can be arranged to contain information. The average information per symbol is called Shannon entropy. It is almost always represented in the literature by a capital H.

H = Shannon Entropy

If all symbols are equally probable, the Shannon entropy is defined as:

$$H = 3.32 \times \log(\text{ number of possible symbols}) \qquad \text{Eq. 1}$$

or for readers with a base 2 log function on their calculator

$$H = \log_2 (\text{number of possible symbols})$$

For example, if 26 blocks are labeled with letters from the alphabet and then drawn from a hat, the average information per symbol is $H = 3.32 \times \log(26) = 4.7$ bits per letter.

Now suppose that 102 blocks with the letter *Z* are added to the hat in the previous example. The hat now contains 128 blocks, but most are labeled with a *Z*. The Shannon entropy is now the weighted average of the contents in the hat. The odds of pulling the letter A from the hat are 1 in 128. So from equation 2, in chapter 1, the information associated with drawing an A is $3.32 \times \log(128/1) = 7$ bits. With the exception of the letter Z, all other letters have the same odds. Thus, observing any letter, A-Y, results in 7 bits of information.

The letter Z will be drawn from the hat 102 times with every 128 tries. This corresponds to 1 time in 1.25 tries, and the corresponding information is 3.32 x log (128/102) = 1.6 bits.

Shannon entropy is the weighted average. 26 symbols contribute 7 bits and 102 contribute 1.6 bits. So H is calculated as follows: (102/128) x 1.6 + 26/128 x (7) = 1.275 + 1.42 = 2.7 bits per symbol. This means that a code exists that can transmit the contents of the hat using on average only 2.7 bits per symbol.

Question: How much information is carried by this message: "AAAAAAAA" if it is drawn from the hat with 128 letters?
Answer: Each A contributes 7 bits, so this message contains 8 letters x 7 bits per letter = 56 bits of information.
Question: what are the odds of drawing it from the hat?
Answer: 1 in 2^{56} or 1 in 7.2×10^{16}.
Question: on average how much information will most 80 letter messages drawn from this hat contain?
Answer: 80 letters x 2.7 bits per letter = 216 bits.
Question: How much information is carried by this message: "ZZZZZZZZ"?
Answer: Each Z contributes 1.6 bits. So this message contains 8 letters x 1.6 bits per letter = 12.8 bits of information.

Shannon entropy only works for very long messages. Short messages may or may not be accurately represented by Shannon entropy. The information content of a short message may be much higher or much lower. The examples above illustrate this concept. Long messages will always converge to the average information per bit. This is why Shannon entropy is useful.

Appendix 2: Relative Entropy and Information

This section is best read immediately after chapter 4. The techniques used in chapter 4 to calculate information are different from those used by most authors. Most use relative entropy.

Suppose that a sequence alignment for a protein gives the followings results (only the first two amino acids in the sequence are shown):

Chicken AlaVal...
Man AlaVal...
Dog AlaVal...
Lizzard AlaAla...
Fish AlaAla...
JellyFish AlaAla...

The information in the first position is easy to calculate. Four out of 64 possible codons specify alanine. So the information is by equation 2 in chapter 1 as follows:

Information = 3.32 x log (64/4) = 4 bits.

The information of the second position is also easy to calculate. Four codons specify alanine and 4 specify valine. The information is as follows:

Information = 3.32 x log (64/ 8) = 3 bits.

These calculations are in full agreement with the techniques described in chapter 4.

Now consider a different alignment.

Chicken	AlaVal..
Man	AlaVal..
Dog	AlaVal..
Lizzard	AlaVal..
Fish	AlaVal..
JellyFish	AlaVal..
Oak Tree	AlaVal..
E coli	AlaAla..

The information content of the first position has not changed. It is still 4 bits, but what about the second position? The same two amino acids are present, but valine is found in 7 of the 8 sequences. Intuitively, it would seem that the second position in this alignment contains more information, and it does. To calculate the information for the second position, the formula for relative entropy must be used. The formula for relative entropy when two amino acids appear in the same alignment column is as follows:

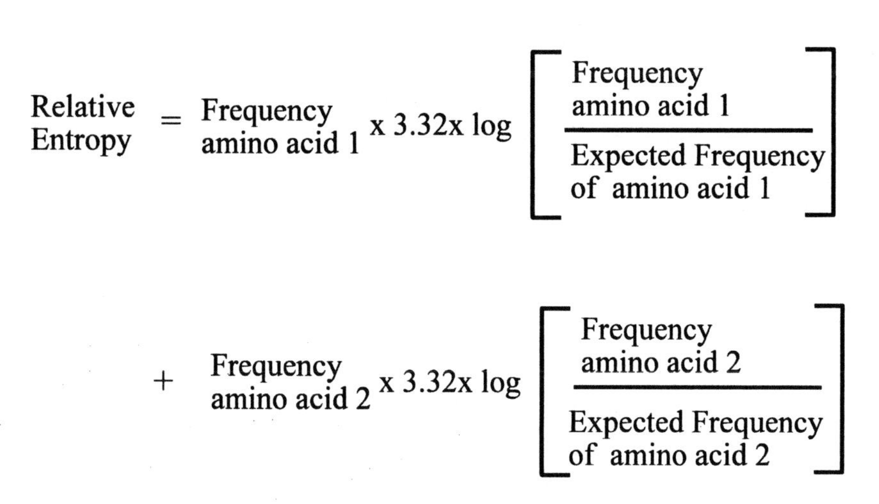

In this case, the relative entropy is as follows:

Relative Entropy =
1/8 x 3.32 x log [(1/8)/(4/64)] + 7/8 x 3.32 x log [(7/8)/(4/64)]

Relative Entropy = .125 + 3.33 = 3.46 bits.

Relative entropy is a measure of information. In fact, the actual information at position 2 is the relative entropy.

Relative Entropy = Actual Information = 3.46 bits.

Now consider what happens when the equation for relative entropy is applied to the second position in the first set of sequences. In this set of sequences, valine occurs ½ of the time, and alanine occurs ½ of the time.

Relative entropy =
½ x 3.32 x log [(½)/(4/64)] + ½ x 3.32 x log[(½) /(4/64)] =
1.5 + 1.5 = 3 bits.

So relative entropy gives the same results as the technique used in chapters 4 and 5 when the amino acids in a column all occur at the expected frequency. The expected frequency is set by the underlying probabilities associated with each amino acid arising by chance.

Why not use relative entropy? Relative entropy is always greater than or equal to the information calculated with the techniques used in this book. These techniques always calculate the minimum possible information in the gene or protein today. The true information which can only be found by using relative entropy will always be higher. Relative entropy depends on the distribution of amino acids, and no such distribution can exist for the very first proteins. So while relative entropy is the correct method to measure information, it is not the best method to model the evolution of new proteins and genes.

Fortunately, with the simple assumption that all amino acids in a distribution occur at the expected frequency, the equation for relative entropy simplifies to those introduced in chapter 1 (pages 24-25). These two equations always calculate the minimum possible information in any distribution. Thus, the technique used in chapters 4 and 5 calculates the minimum possible information by assuming that all allowed amino acids in any given column are found at the expected frequency. Under some circumstances, this value can be related to a probability for evolution (chapter 5). But most of the time, this should never be done. Molecular knowledge defines the minimum information required for a selective advantage to be realized. It can almost always be related to a probability for evolution. The size of this first vertical step determines whether or not chance will create the required knowledge. The graph below shows how these quantities are related.

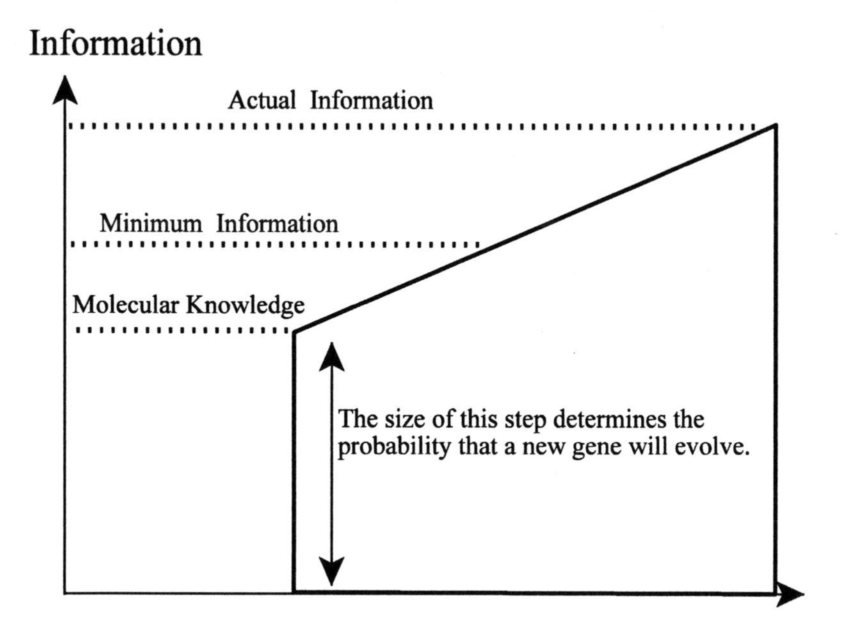

Appendix 3: Math Review

Base 10 is composed of 10 integers. These are 0,1,2,3,4,5,6,7,8, and 9. Using these 10 integers, any number can be expressed quite easily. Base 10 is called base 10 because it uses 10 unique digits to express numbers. Most math problems use base 10.

Base 2 is composed of 2 integers. These are 1 and 0. Just like with base 10, any number can be expressed in base 2. Computers use base 2 for mathematical calculations. The following table shows how to count from 0 to 7 in both systems.

Base 10	Base 2
0	000
1	001
2	010
3	011
4	100
5	101
6	110
7	111

Base 2 is very useful for information theory because each digit in base 2 contains exactly 1 bit of information. So when information was defined, engineers naturally decided to define it in terms of base 2.

Exponents

Exponents indicate that a number should be multiplied by itself a certain number of times. In the following equation: $10^3 = 10 \times 10 \times 10 = 1000$, the exponent is 3 and in this equation $2^4 = 2 \times 2 \times 2 \times 2 = 16$, the exponent is 4. Any number multiplied by itself zero times is by definition equal to 1, so $10^0 = 2^0 = 10{,}000^0 = 1$.

Logarithms

The symbol log on a calculator indicates a logarithm. Logarithms are the inverse operation of exponents. For example, if the exponent 3 is used to raise 10 to a power the result is 1,000. If the log function is used to find the logarithm on 1,000, the result is equal to the exponent which in this case is 3.

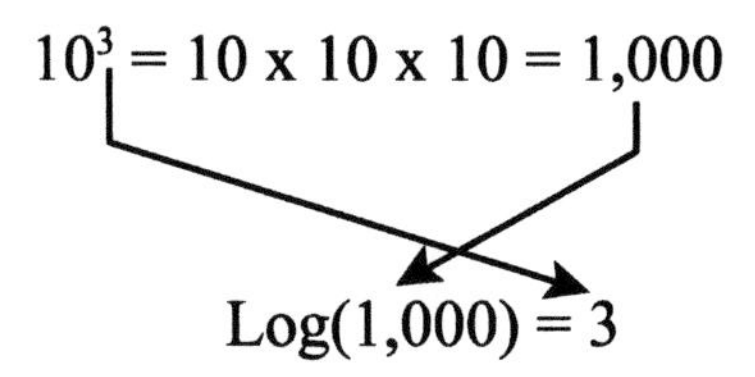

Logarithms must assume a base. The log function on a calculator assumes that the user wants to calculate the logarithm for base 10. The equation below does not work because the log function is only the inverse when the number 10 is raised to a power by the exponent.

$$2^3 = 2 \times 2 \times 2 = 8$$

$$\mathrm{Log}(8) = 0.9 \neq 3$$

Because information theory uses base 2, the exponents always operate on the number *2*. The log function can be used to find the inverse of this operation if the answer is multiplied by a conversion factor, 3.32. This only works for exponents raising the number *2* to a power.

$$2^3 = 2 \times 2 \times 2 = 8$$

$$3.32 \times \mathrm{Log}(8) = 3$$

Addition with logarithms is the same operation as multiplication. The following equation will multiply 100 x 100 using logarithms.

$\text{Log}(100) + \text{Log}(100) = 2 + 2 = 4$

The number *4* is now used to raise the number *10* to a power:

$10^4 = 10 \times 10 \times 10 \times 10 = 10{,}000$

Notice that 100 x 100 also equals 10,000. Today with handheld calculators logarithms are seldom used to multiply large numbers, but the property of logarithms that turns multiplication into addition is still very useful for information theory. It is this property of logarithms that led scientists to define information using logarithms because it allows information to be added without having to worry about multiplication.

Consider the following two sentences:

I like the dog.

I like the cat.

Both sentences use 15 characters, 11 letters, 3 spaces, and 1 period. If the sentences are constructed by drawing blocks labeled with the 26 letters in the alphabet, 1 space and 1 period, the odds of drawing either sentence is 1 in 5.1×10^{21}. The odds of drawing the first sentence followed by the second are 1 in $(5.1 \times 10^{21}) \times (5.1 \times 10^{21})$ = 1 in 2.6×10^{43}. Intuitively, these two sentences together should have twice as much information as either one by itself, and this is where the log function comes into play.

$3.32 \times \text{Log}(5.1 \times 10^{21}) = 72$ bits for either sentence alone.

$3.32 \times \text{Log}(2.6 \times 10^{43}) = 144$ bits for both sentences

Appendix 4: A Review of Yockey's Approach

Today, many scientists apply information theory to molecular biology, but only a few have tried to use information theory to answer the most important question. Is evolution possible? Yockey was probably the first. Because so few scientists are trying to answer this question, a consensus as to how to assign information has not been reached.

Yockey assigns information to proteins by a technique that is very different from the one developed in chapter 4. His technique relies on both Shannon entropy and conditional entropy. He models the transfer of information from DNA to proteins and assigns information to this transfer by treating mutations as a source of noise. Unfortunately, his analysis fails to consider natural selection, and so the information that he calculates is incorrect. Natural selection weeds out harmful mutations. In a communication system, this is analogous to a person receiving an unintelligible message and then asking for the message to be re-sent. The equations that Shannon developed to model noise in communication systems do not apply if a person on the receiving end inspects the incoming messages for errors and discards any messages that contain errors.

This book did not use Yockey's technique because the information assigned by his technique cannot be related to a probability for protein evolution. For a full development of his technique, refer to the two references at the end of this appendix.

Yockey's technique divides every sites in a protein into two categories, absolutely conserved and not absolutely conserved. An absolutely conserved amino acid is one that never changes. For example, if column 20 in a multiple sequence alignment is always a glycine, then position 20 in the alignment is absolutely conserved. If more than one amino acid is found in column 20, then it is not absolutely conserved. Yockey's technique sets the information content of any absolutely conserved amino acid equal to the Shannon entropy, 4.14 bits.

So using this technique, a peptide composed of 10 methionines has the same chance of evolving as one composed of 10 serines. This is the drawback of using Shannon entropy. Methionine is only specified by 1 codon, and serine is specified by six. Assuming random mutations, serine should arise by chance six times as often as methionine.

For reference, the Shannon entropy, assuming the genetic code, is calculated below. To follow Yockey's method exactly only the 61 codons that do not terminate the peptide chain are allowed.

amino acid group	expected frequency	information	total (bits)
6 codons (3 amino acids)	29.5%	3.34	0.99
4 codons (5 amino acids)	32.8%	3.93	1.29
3 codons (1 amino acid)	4.9%	4.34	0.213
2 codons (9 amino acids)	29.5%	4.93	1.45
1 codon (2 amino acids)	3.3%	5.93	0.195
			4.14

Example calculation: Three amino acids are specified by 6 codons. The information acquired when one of these is observed is as follows: information = 3.32 x log (61/6) = 3.34 bits. Because these amino acids should arise by chance 6 times in every 61 tries and because there are 3 of these amino acids, the expected frequency is (3 x 6)/61 = .295 or 29.5%. The last column is the product of the expected frequency and the information for each amino acid group (each row). The sum of all entries in the last column is the Shannon entropy.

Because proteins are short messages, Shannon entropy is almost never a true measure of a specific protein's information. A typical protein may only have 30 absolutely conserved amino acids, and the probability for these amino acids arising by chance might be very different from the number calculated using Shannon entropy.

Yockey's calculations for amino acids that are not absolutely conserved run into another issue. He models mutations as noise, and then applies the techniques developed by Shannon to calculate information transfer through a noisy communication channel. The figure below illustrates the similarity between information transfer in life and in electrical communication systems.

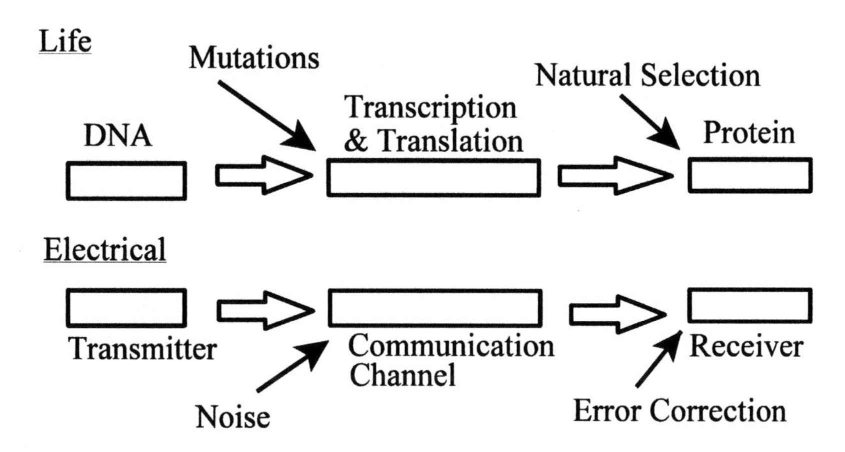

In electrical systems, conditional entropy models the amount of information lost due to noise. For example, suppose the transmitter transmits the results of a trapped scientist experiment. The two results are heads or tails. If the channel is noisy, when heads is transmitted tails might be received. This error reduces the rate of information transfer. Conditional entropy models how much information is lost. Mutations in life have the same effect as noise in communication systems, so mutations can be modeled as a noise source.

The problem with this approach is that natural selection does not allow harmful mutations to survive. So if a mutation creates a non-functional protein, the mutation will be removed from the population by natural selection. In communication systems, error correction is responsible for this function. With error correction, conditional entropy can no longer be used to accurately model the information lost due to noise. Likewise, because of natural selection, conditional entropy does not model the effect of mutations on information transfer.

Natural selection cannot ensure that only allowed messages are transmitted, but it does ensure that only allowed messages survive. And the neutral theory of evolution (see chapter 4) predicts that many if not all of the allowed amino acid substitutions in the final protein will be observed if the same proteins in many diverse species are analyzed. Thus, the information content of the final protein does not depend on the information transferred from DNA, and conditional entropy is not needed for this analysis. The equations to calculate information in chapter 1 are applicable, and they may be applied using the techniques introduced in chapters 4 and 5.

Furthermore, the techniques used in this book always maintain a strict one to one relationship between probability space and information space. In other words, probability theory and information theory both yield the same results. Yockey's approach does not preserve this one to one mapping. So the odds that he calculates for protein evolution are different than the odds that one would calculate using probability theory.

References:

1) Yockey, Information Theory, Evolution and the Origin of Life, 2005.

2) Yockey, Information Theory and Molecular Biology, 1992.

3) Shannon, A Mathematical Theory of Communications, 1948.

Glossary:

Acid - an acid is a chemical that lowers the pH of a solution. Acids interact with other chemicals by donating hydrogen atoms or accepting electrons.

Activation Energy - an energy barrier that must be crossed before chemicals can react.

Actual Knowledge - the molecular knowledge contained in a gene or protein.

Adenine - one of the bases in DNA and RNA. It is also one of the components in ATP, ADP and AMP.

ADP - adenine diphosphate. ADP only has one high energy bond.

Alanine - hydrophobic amino acid.

Amino Acid - the building blocks of proteins. Life uses 20 amino acids to build proteins. Amino acids have two sticky ends that can be joined together to form long chains.

AMP - adenine monophosphate. AMP has no high energy bonds. It is created in the cell from ATP when ATP is used to accomplish some task that requires energy.

Arginine - a basic amino acid.

Aspartate - polar acidic amino acid.

Asparagine - charged amino acid.

ATP - adenine triphosphate. This is the energy source used by life. It is made by plants during the process of photosynthesis. It is made by animals as they digest food. ATP has 2 high energy bonds. Life knows how to use the high energy bonds to do work.

ATP synthase - an enzyme that synthesizes ATP.

Axiom - a self evident assumption.

Base - a base is a term used in chemistry to describe a chemical that raises the pH of a solution. Bases interact with other chemicals by accepting hydrogen atoms or donating electrons.

Bit - a unit of information or knowledge.

Calculated knowledge - molecular knowledge calculated by the methods outlined in chapter 4.

Carboxylic acid - one of the sticky ends on an amino acid.
Cell - the smallest living unit. Cells are surrounded by a membrane which is composed of lipids and proteins. The chemicals inside cells know how to grow and replicate. They accomplish this by using an abundant energy source to do useful work.
Chemical evolution - the hypothetical process by which the chemicals necessary for life emerged on the primitive earth.
Chemical oscillators - a system of chemicals that change in a periodic fashion with time. A wrist watch is a chemical oscillator.
Chromosome - very large continuous molecules of DNA.
Clustal - a computer program used to align amino acids in proteins.
Codon - a grouping of 3 bases in DNA or RNA that specifies a specific amino acid in the final protein.
Common Ancestor - a animal, plant or bacteria whose descendants evolved into two or more species.
Complexity - a non-repetitive pattern.
Condensation agent - a chemical that facilitates the formation of chemical bonds between biological molecules by absorbing water.
Creation science - belief that the biblical account of creation is scientifically accurate.
Cysteine - amino acid that can cross link peptide chains.
Cytosine - one of the bases used in DNA and RNA.
C-terminus - one of the sticky ends of an amino acid.
Darwinian evolution - evolution that involves small steps in molecular knowledge. Naturalistic laws explain Darwinian evolution.
Deoxyribose - the sugar molecule found in DNA.
DNA - stands for deoxyribose nucleic acid. DNA is the molecule that stores all of the knowledge that life needs to grow and replicate. Sections of DNA that contain the knowledge to build proteins are called genes.
Energy - the ability to do work.
Entropy - a measure of uncertainty.
Enzyme - a protein that catalyzes a chemical reaction.

Evolution by design - evolution that involves gigantic steps in molecular knowledge. Naturalistic laws do not explain evolution by design.

Evolution - the process through which existing knowledge is optimized and preserved. While science relies of evolution to create new knowledge, the probabilities associated with these events are too remote.

Eukaryote - cells with a defined nucleus. Plants and animal cells are composed of eukaryotic cells.

G3PD - an enzyme used by life to metabolize sugar.

Gene duplication - the process by which existing genes are duplicated.

Gene - a section of DNA that contains the knowledge to build a protein.

Genetic Code - the code that is used to build proteins from the knowledge contained in DNA.

Glutamate - polar acidic amino acid.

Glutamine - polar charged amino acid.

Glycine - a small amino acid.

Guanine - one of the bases found in DNA and RNA.

Heat - the flow of energy from hot to cold objects.

Histidine - a basic amino acid.

Hydrophobic - a molecule that does not like water.

Knowledge - a useful reduction in uncertainty.

Leucine - hydrophobic amino acid.

Life - a system of chemicals possessing the molecular knowledge and a mechanism to implement this knowledge in such a way to survive long enough to replicate.

Lysine - a basic amino acid.

Infon - a step in molecular knowledge found by chance.

Information - a reduction in uncertainty. Information is not necessarily useful.

Insulin - a protein that signals cells to take up sugar from the blood stream.

Intelligent design - a methodology that relies on indirect logic to infer the existence of a creator.

Investigator interference - process by which researchers alter the results of their experiments.

Irreducible complexity - a system that requires two or more components to function.

Isoleucine - hydrophobic amino acid.

Meteorites - rocks that fall from space and are recovered. Some contain organic chemicals like amino acids.

Methionine - hydrophobic amino acid.

Micro-state - the arrangement of particles in a system.

Molecule - a chemical composed of two or more atoms.

Molecular Knowledge - the minimum amount of information necessary to enable a chemical or group of chemicals to accomplish some task or to specify some trait.

mRNA - stands for messenger ribonucleic acid. RNA is an intermediate molecule involved in protein synthesis. Messenger RNA transfers the knowledge to build proteins from the nucleus to the ribosomes.

Natural selection - the process by which nature ensures that only optimized genes are passed onto future generations. Natural selection works against evolution because it does not allow existing genes to evolve.

Naturalistic axiom - the assumption that everything can be explained by physics, chemistry, biology and math.

N-terminus - one of the sticky ends of an amino acid.

Nucleotide - the building block for DNA and RNA. Each nucleoide contains 1 phosphate, 1 ribose, and one of the 5 bases, adenine, guanine, cytosine, thymine, or uracil.

Nucleus - the central compartment in eukaryotic cells. The nucleus contains the DNA.

Peptide - a short chain of amino acids.

Perpetual motion machines - a machine that violates one or more of the laws of physics.

Phenylanlanine - bulky amino acid.

Phosphodiester bond - the high energy bond between phosphate groups in ATP.

Polar - a molecule that likes water.

Poly-nucleotide - a string of nucleotides.

Pre-RNA - hypothesized molecule that proceeded RNA during chemical evolution.

Primordial soup - the hypothetical small pond in which life originated.

Primordial information - the knowledge required to exclude chemicals in the primoridal soup from growing biological molecules.

Prokaryote - cells without a defined nucleus. Bacteria cells are prokaryotic cells.

Proline - unusual amino acid that introduces a kink into a protein chain.

Protein - a chemical composed of amino acids that is used by life to implement the molecular knowledge found in genes.

Protein domain - a section of a protein that performs some function or specifies some 3-D shape.

Protein family - a group of related proteins with similar 3-D structures and amino acid sequences.

Proteinoids - long branched chains of amino acids formed by heating.

Observable axiom - the assumption that man is capable or accurately observing the world around him.

Order - a repetitive pattern.

Oxidation - a chemical reaction in which electrons are transferred from one atom to another.

Quantum mechanics - the physics that describes small particles.

Rasmol - a free computer program that allows users to view molecules like proteins and DNA.

Relative entropy - a measure of uncertainty.

Ribose - the sugar molecule found in RNA.

Ribozyme -an RNA molecule that also functions as an enzyme.

Ribosome - a complex of RNA and proteins. Proteins are built by the process of translation at ribosomes.

RNA - RNA contains molecular knowledge and can sometimes implement molecular knowledge. Messenger RNA (mRNA), transfer RNA (tRNA), and ribosomal RNA (rRNA) help cells use the knowledge contained in DNA to build proteins.

Second law of thermodynamics - states that the entropy of the universe increases with time. This happens because particles always try to find their most probable distribution. This is the distribution that maximizes the number of available micro-states.

Serine - a polar amino acid.

Science - a methodology that relies on the observable and naturalistic axioms to design and interpret scientific experiments.

Shannon entropy - the average uncertainty per symbol.

Species - a group of interbreeding animals.

Specified Complexity - any outcome that is not ordered and is predicted in advance.

Swiss Prot - an online Database that contains the amino acid sequence of most proteins.

Thermal proteins - long branched chains of amino acids formed by heating.

Thermodynamics - the field of physics that deals with heat, energy and work. Chemical thermodynamics is a subset of this discipline that deals with chemicals and how they interact.

Threonine - a polar amino acid.

Transcription - the process that writes DNA into mRNA.

Translation - the process that uses mRNA to create proteins.

Thymine - one of the bases found in DNA.

tRNA - stands for transfer ribonucleic acid. Transfer RNA brings amino acids to ribosomes for protein synthesis.

Tyrosine - bulky amino acid.

Tryptophan - bulky amino acid.

Uracil - one of the bases found in RNA.

Valine - hydrophobic amino acid.

This book would not have been possible without numerous online programs and resources.

The Protein Database:
www.rcsb.org/pdb/ The protein database that houses the 3-D structure of many proteins. These can be viewed online with interactive programs that allows users to zoom in and rotate the structures.

Rasmol:
www.umass.edu/microbio/rasmol/distrib/rasman.htm Displays 3-D structures of proteins in the protein database. Used to create most of the pictures inside this book and on the cover.

Swiss Prot:
www.us.expasy.org/sprot/ Contains the amino acid sequence of most proteins.

Consurf:
www.consurf.tau.ac.il/ Colors 3-D displays by amino acid conservation. Generates a script to apply in rasmol. This program was used to color the amino acids on the back cover.

Glaser F./ Pupko T., Paz I., Bell R.E., Becher D., Martz E., Ben-Tal N., Consurf: Identification of Functional Regions in Proteins by Surface Mapping of Phylogentic Information, Bioinformatics, Vol. 19, no, 1, 2003 pp163-164.

Clustal:
http://www-igbmc.u-strasbg.fr/BioInfo/ClustalX/Top.html An easy to use windows interface for the Clustal alignment program. This program was used in chapter 4 to align insulin in several different species and again in chapter 14 for G3PD.

Also this book has a companion web site which can be found at this URL address: www.evolution-by-design.com. The password for the pdf documents is molecular-knowledge.

Subject Index:

Author Index:

WITHDRAWN

JUN 0 5 2024

DAVID O. McKAY LIBRARY
BYU-IDAHO

Printed in the United States
36845LVS00001B/5